THE STORY OF EVOLUTION IN 25 DISCOVERIES

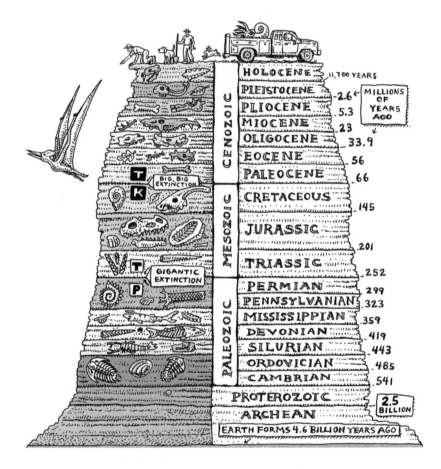

HOLOCENE ··········· 11,700 YEARS

PLEISTOCENE ····· 2.6 ←

MILLIONS OF YEARS AGO

PLIOCENE ········· 5.3

MIOCENE ·········· 23

OLIGOCENE ········· 33.9

EOCENE ············ 56

PALEOCENE ········· 66

CENOZOIC

CRETACEOUS ········· 145

JURASSIC ··········· 201

TRIASSIC ··········· 252

MESOZOIC

BIG, BIG EXTINCTION

GIGANTIC EXTINCTION

PERMIAN ··········· 299

PENNSYLVANIAN ······ 323

MISSISSIPPIAN ······· 359

DEVONIAN ·········· 419

SILURIAN ··········· 443

ORDOVICIAN ········· 485

CAMBRIAN ·········· 541

PALEOZOIC

PROTEROZOIC

ARCHEAN

2.5 BILLION

EARTH FORMS 4.6 BILLION YEARS AGO

THE STORY OF EVOLUTION

n 25 DISCOVERIES

THE EVIDENCE AND THE PEOPLE
WHO FOUND IT

DONALD R. PROTHERO

COLUMBIA UNIVERSITY PRESS   NEW YORK

COLUMBIA UNIVERSITY PRESS

Publishers Since 1893

New York   Chichester, West Sussex

cup.columbia.edu

Copyright © 2020 Donald R. Prothero

Library of Congress Cataloging-in-Publication Data

Names: Prothero, Donald R., author.

Title: The story of evolution in 25 discoveries : the evidence and the people
   who found it / Donald R. Prothero.

Description: New York : Columbia University Press, [2020] |
   Includes bibliographical references and index.

Identifiers: LCCN 2020013001 (print) | LCCN 2020013002 (ebook) |
   ISBN 9780231190367 (hardback) | ISBN 9780231548854 (ebook)

Subjects: LCSH: Evolution—Research. | Evolution—History.

Classification: LCC QH362 .P76 2020 (print) | LCC QH362 (ebook) | DDC 576.8072—dc23

LC record available at https://lccn.loc.gov/2020013001

LC ebook record available at https://lccn.loc.gov/2020013002

∞

Columbia University Press books are printed on permanent and durable acid-free paper.

Printed in the United States of America

Cover design: Elliott S. Cairns
Cover illustration: Trudy Nicholson
Frontispiece: Courtesy of Ray Troll

• ◉ •

THIS BOOK IS DEDICATED TO THE MEMORY OF MY FRIEND
AND MENTOR, STEPHEN JAY GOULD (1941-2002).
HE INSPIRED MY ENTIRE GENERATION OF PALEONTOLOGISTS
AND EVOLUTIONARY BIOLOGISTS WITH HIS GREAT RESEARCH
AND INSIGHTS INTO IMPORTANT TOPICS.
HE DID SO MUCH TO EDUCATE THE GENERAL PUBLIC ABOUT
THE REALITIES OF EVOLUTION.
HE SHOWED THAT NATURAL HISTORY WRITING COULD
BE ENGAGING AND EXCITING, AND THAT THE PUBLIC
COULD UNDERSTAND COMPLEX IDEAS IF THEY ARE
EXPLAINED IN A CLEAR AND ENTERTAINING WAY.

There is grandeur in this view of life, with its several powers,
having been originally breathed into a few forms or into one; and that,
whilst this planet has gone cycling on according to the fixed law of gravity,
from so simple a beginning endless forms most beautiful and
most wonderful have been, and are being, evolved.

—CHARLES DARWIN, 1859

Nothing in biology makes sense except in the light of evolution.

—THEODOSIUS DOBZHANSKY, 1973

CONTENTS

PREFACE   XI
ACKNOWLEDGMENTS   XIII

 IN THE BEGINNING
EVERYTHING EVOLVES,
AND EARTH IS VERY OLD

 1 EVERYTHING EVOLVES
AND CHANGES   3
DISCOVERY OF THE EVOLVING UNIVERSE

 2 THE ABYSS OF TIME   16
THE IMMENSE AGE OF THE EARTH

 II DARWIN'S EVIDENCE
FOR EVOLUTION

3 EVOLUTION IN ACTION   31
TRANSFORMATION IN REAL TIME

 4 OUR COMMON BODY PLAN   47
HOMOLOGY

 5 ONTOGENY RECAPITULATES
PHYLOGENY   61
EVIDENCE IN EMBRYOS

 6 THE SINKING OF NOAH'S ARK   72
BIOGEOGRAPHY

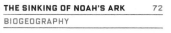

 7 THE BRANCHING TREE OF LIFE   86
PHYLOGENY

 8 THE CASE OF THE
CRUEL WASPS   97
NATURE IS NOT MORAL

 9 JURY-RIGGED CONTRIVANCES   108
NATURE IS NOT OPTIMALLY DESIGNED

III **GREAT TRANSITIONS IN THE HISTORY OF LIFE**

10 **A WHALE OF A TALE** 121
VESTIGIAL ORGANS AND WALKING WHALES

11 **INVASION OF THE LAND** 136
AMPHIBIANS CRAWL OUT OF THE WATER

12 **MISSING LINKS FOUND** 150
MACROEVOLUTION AND TRANSITIONAL FOSSILS

13 **BIRDS WITH TEETH** 168
THE DINOSAURS AMONG US

14 **A HORSE! A HORSE! MY KINGDOM FOR A HORSE!** 179
THE EVOLUTION OF EQUINES

15 **HOW THE GIRAFFE GOT ITS NECK** 195
LAMARCK, DARWIN, AND THE LEFT RECURRENT LARYNGEAL NERVE

16 **HOW THE ELEPHANT GOT ITS TRUNK** 207
THE EVOLUTION OF PROBOSCIDEANS

IV **EYES AND GENES**

17 **A WARM LITTLE POND** 225
HOW DID LIFE ORIGINATE?

18 **GENETIC JUNKYARD** 241
MOST OF OUR DNA IS USELESS

19 **LEGS ON THEIR HEADS** 253
HOMEOTIC MUTANTS AND EVO-DEVO

20 **THE EYES HAVE IT** 260
THE EVOLUTION OF PHOTORECEPTION

V **HUMANS AND EVOLUTION**

21 **A TINKERER, NOT AN ENGINEER** 273
ARE HUMANS WELL DESIGNED?

22 **THE THIRD CHIMPANZEE** 289
ARE WE REALLY 99 PERCENT THE SAME?

23 **THE APE'S REFLECTION** 297
ARE HUMANS REALLY THAT DIFFERENT FROM OTHER ANIMALS?

24 **BONES OF OUR ANCESTORS** 311
THE HUMAN FOSSIL RECORD

25 **THE ONCE AND FUTURE HUMAN** 332
ARE HUMANS STILL EVOLVING?

INDEX 345

PREFACE

Since the publication of my book *Evolution: What the Fossils Say and Why It Matters* (Columbia University Press, 2007; second edition, 2017), the fields of evolutionary biology and paleontology have made many new discoveries. Meanwhile, the evidence for evolution has been piling up and accumulating since publication in 1859 of the revolutionary book, *On the Origin of Species*, by Charles Darwin. Some of that evidence is discussed in my first evolution book, but much of it is new, or mentioned only briefly in that book.

Rather than focusing exclusively on the fossil record and spending a lot of time correcting the lies and myths of the evolution deniers, I thought it would be interesting and useful to focus on individual lines of evidence that led to the discovery of evolution, and the powerful insights they give us into the way that life works. Mara Grunbaum's *WTF, Evolution: A Theory of Unintelligible Design* (2014) and other books have made the point that life is full of bizarre and funny and ugly things that make no sense in a divinely designed universe, showing how clumsy and wasteful nature can be. However, that is basically a picture book full of jokes and one-liners with a hip, irreverent attitude. I want to make the same point in a more serious way, exploring this topic and delving deeper into its meaning.

In addition, I enjoy writing in the format of the three previous books in this series, *The Story of Life in 25 Fossils* (Columbia University Press, 2015), *The Story of the Earth in 25 Rocks* (Columbia University Press, 2018), and *The Story of Dinosaurs in 25 Discoveries* (Columbia University Press, 2019). Each chapter in this book, as in the previous titles, is a self-contained vignette describing one particular idea, often wrapped in the historical context of how people have thought about this topic. As in the previous books, the science is often framed in terms of the stories of the people who made the discoveries, and the importance of the discovery in the context of science.

The book is organized into five sections. Part I, "In the Beginning: Everything Evolves, and Earth Is Very Old," describes how evolution is happening throughout the universe, which is billions of years old (chapters 1–2). Part II discusses Darwin's original main lines of evidence for evolution, and the related discoveries that have happened since Darwin's time (chapter 3–9). Part III, "Great Transitions in the History of Life," talks about the dramatic evidence from the fossil record illustrating how certain major groups of organisms evolved from something completely different, or macroevolution (chapters 10–16). Part IV, "Eyes and Genes," describes the enormous volume of evidence from genetics and molecular biology (chapters 17–19), and it also deals with the famous conundrum of how a complex structure like the eye could evolve (chapter 20). Part V, "Humans and Evolution," details the evidence supporting the idea that humans are apes and evolved, much like any other organism (chapters 21–24) and where future evolution might go—and will not go (chapter 25).

Like Darwin did in 1859, I hope to convince you, the reader, of the reality of evolution by building the case one anomalous fact of nature at a time. Each is clear evidence of evolution, and I hope you will be persuaded of the wonders of evolution just by the sheer overwhelming weight of the evidence. This contrasts with my first evolution book, which was more scholarly and philosophical and dealt directly with creationism and the broader topics around evolution.

So sit down and prepare to explore the wondrous and bizarre aspects of nature that show how it has a history, and how it evolved and changed through time. As the great Theodosius Dobzhansky wrote in a 1973 essay, repeating these words in his title, "Nothing in biology makes sense except in the light of evolution."

ACKNOWLEDGMENTS

I thank my former editor at Columbia University Press, Patrick Fitzgerald, for urging me to write this book and getting the process started, and my current Columbia editor, Miranda Martin, for seeing it to fruition. I thank Julia Kushnirsky and Elliott Cairns at Columbia University Press and Ben Kolstad at Cenveo for their tireless work on producing the book. I thank three anonymous reviewers for their early input on the project, and Pat Shipman and Norman Johnson for reviewing the finished book, or at least some chapters. Finally, I thank my supportive family—my sons Erik, Zachary, and Gabriel, and my amazing wife, Dr. Teresa LeVelle—for tolerating my long absences working at the computer while I wrote this book during the summer of 2019.

THE STORY OF EVOLUTION IN 25 DISCOVERIES

PART I

IN THE BEGINNING

EVERYTHING EVOLVES, AND EARTH IS VERY OLD

Figure 1.1 ▲

A famous engraving from an 1888 book by the French astronomer Nicholas Camille Flammarion showing the medieval conception of Earth as a flat disk surrounded by the fixed stars on a celestial sphere. The curious explorer pokes his head through the "dome of the sky" to see the sun, moon, and planets moving on great gear wheels as they orbit around us. (Courtesy of Wikimedia Commons)

EVERYTHING EVOLVES AND CHANGES

There's nothing constant in the Universe,
All ebb and flow, and every shape that's born
Bears in its womb the seeds of change.

—OVID, *METAMORPHOSES* XV (8 CE)

πάντα χωρεῖ καὶ οὐδὲν μένει
Everything changes and nothing stands still.

—HERACLITUS, CA. 500 BCE, AS QUOTED BY PLATO IN *CRATYLUS*

Our view of the universe and the solar system has changed dramatically in the past 500 years. Before 1543, almost all humans thought the earth was flat and was at the center of the universe and that the stars were tiny points of light on the dome of the heavens (figure 1.1). In 1543, Copernicus provided evidence for the idea that the sun, not Earth, was at the center of our world, and that Earth was a planet in orbit around the sun. In 1609, Galileo used a newly invented device called a telescope to discover that the stars were beyond counting and that they were not scattered on a big dome over our heads. He also confirmed that Jupiter has its own moons that could move completely around it, showing that it was not sitting in a perfect celestial sphere or dome above Earth. He debunked the notion that the planetary bodies were perfect and unsullied when his telescope revealed that Earth's moon is covered in craters and is not a perfect celestial sphere. Finally, by discovering that Venus has phases similar to those of Earth's moon ("quarter Venus," "half Venus," etc.), he showed that Venus was moving around the sun in an orbit inside our own orbit. Most important, he confirmed Copernicus's idea that Earth was just another planet orbiting the sun. By the 1670s and 1680s, Isaac Newton had worked out the laws of motion and gravitation, illustrating how the entire system could be explained by basic physics.

Today we look at the amazing images of space coming from both land-based telescopes and the Hubble Space Telescope, and we see what no one could have possibly imagined even 30 years ago. We can see the stages of how stars are born and die and how other planets and solar systems have formed. These images, and the astrophysical calculations and models that explain them, give us a new view of the origin of the solar system and allow us to explain much of what was simply guesswork before this century.

Where did we come from? When and where did it all begin? These questions have fascinated and troubled people since humans first looked at the skies. For millennia, the explanations came from a wide variety of religious myths and stories representing every culture on Earth. Early in the twentieth century it became possible to go beyond myth and speculation, and we began to use the methods of science to discover what really happened.

The first breakthrough came from a number of women astronomers (figure 1.2) working at Harvard College Observatory under W. C. Pickering. They were known as the "Harvard Computers" because they were talented mathematicians who could quickly make calculations and computations in their head and on paper and do measurements by hand. (Only much later did the word "computer" come to mean the electronic devices we all use.) Pickering hired them because they were not only good at math but also careful and meticulous in studying and analyzing thousands of glass photographic plates of the night sky shot by different telescopes. They were also cheaper than male assistants (25 cents an hour, less than a secretary) and worked hard, without complaining, six days a week. This was a time when most women were barred from scientific careers completely, and those who tried to get an advanced education in science met huge barriers every step of the way.

However, their talents soon emerged, and they each made discoveries that revolutionized astronomy and outshone most other male astronomers

Figure 1.2 ▲

The "Harvard Computers" at work around 1890. Henrietta Swan Leavitt is seated third from left with the magnifying glass. To the right of her is Annie Jump Cannon, with Williamina Fleming standing, and Antonia Maury at the extreme right. (Courtesy of Wikimedia Commons)

of their time. The most famous was Annie Jump Cannon, who catalogued the stars of the night sky and proposed the first system of star classification, which was based on their temperatures. She built upon the first complete star classification system by Antonia Maury.

For our story, however, the key woman was Henrietta Swan Leavitt. She was assigned to study "variable stars," classes of stars whose brightness fluctuated from one night to the next. She soon realized that their brightness variations had a regular period of fluctuation, with the brightest stars (most luminous stars) having the longest periods of brightness variation. She found variable stars in a cluster in the constellation Cepheus (thus known as "Cepheids") that were all the same distance away, which allowed her to calibrate the brightness spectrum. In 1913, after studying

1,777 variable stars, Leavitt worked out the relationship between the period of brightness fluctuation and the luminosity of these stars, which enabled her to determine how far away from us a star was by measuring its luminosity and its period of fluctuation. Thanks to Leavitt, astronomers now had a reliable tool to measure how far away a star or galaxy was from Earth.

The next step was made by the legendary astronomer, Edwin Hubble. In 1919, he was assigned to work at the newly completed Mount Wilson Observatory (figure 1.3A) in the mountains above Pasadena, California, and he had free use of what was then the world's most powerful telescope, a reflecting telescope with a 100-inch mirror (figure 1.3B; figure 1.3C). His first major discovery in 1924 used Leavitt's Cepheid variable stars to show that what astronomers had known as "spiral nebulae" were in fact galaxies outside our own Milky Way galaxy and that the Milky Way was just one of many galaxies. This expanded our understanding of the size of the universe far beyond what people had once thought possible.

Hubble used the telescope to systematically study as many stars and galaxies and other large celestial objects as he could. He measured their distance using the Cepheid variable method, and he also used the work of Dutch astronomer Vesto Slipher at Lowell Observatory in Flagstaff, Arizona, who analyzed spectra of light from stars. Just as a prism splits sunlight into its major colors, the light from stars also can be split into a spectrum of colors (figure 1.4). This spectrum, however, has distinctive "bands" across the color scale that are caused by the absorption of certain elements. When we analyze the spectrum of burning sodium or other metals in the lab, we find these same bands, and we can identify the elements we are seeing in each set of bands.

Hubble's major collaborator in this effort was Milton Humason, who had no education beyond age 14 but was eager to prove himself. Humason originally drove the mules that hauled the telescope and other materials up that steep mountain. He then became a janitor during the night shift when the astronomers were at work, so Hubble got to know him. He found that Humason had unexpected talents and promoted him to be his assistant. Hubble admired Humason's quiet determination to take the difficult photographs and to do the careful measurements of the spectrum of thousands of photographic plates from the telescope.

After measuring hundreds of different stars and galaxies, Hubble and Humason noticed something peculiar. The nearest stars had absorption

Figure 1.3 ▲ ▶

(A) Edwin Hubble at the main telescope on Mount Wilson. (B) The largest of the three domes on Mount Wilson, which houses (C) the 100-inch reflecting telescope that Hubble used. ([A] Courtesy of Wikimedia Commons; [B, C] photographs by the author)

Figure 1.3 ▲
(*continued*)

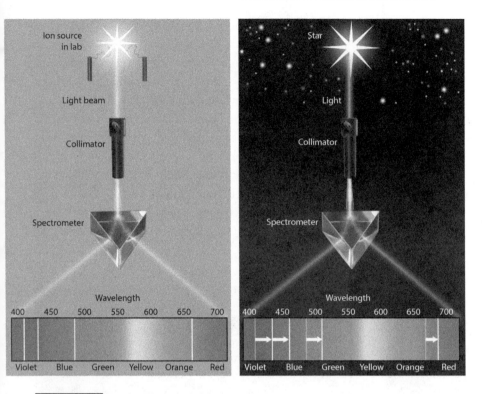

Figure 1.4 ▲

When the spectrum of starlight is broken up through a prism, the different colors and wavelengths are revealed alongside white absorption bands that indicate different elements such as sodium and calcium in the spectrum at different wavelengths. Light from distant stars shows absorption bands that are shifted to the red end of the spectrum compared to their normal positions as determined by a source in the lab. (By permission of Oxford University Press)

lines in their spectra that resembled the same spectrum for that element on Earth. But the farther away the star or galaxy was from Earth, the more the white absorption bands were shifted from their original position and toward the red end of each spectrum.

Why do the absorption lines move toward the red end of the spectrum? This discovery had first been reported and explained for a few galaxies in 1912 by Vesto Slipher. It is known as a Doppler shift and is caused by the Doppler effect. You have experienced the Doppler effect for sound many times. For example, if you are standing on the street when a car or train rushes toward you blaring its horn, you will notice that the pitch of the sound gets slightly higher as it approaches. Once the vehicle has passed you and is rushing away from you, you will hear the sound of the horn drop in pitch again. The Doppler effect is caused by the fact that the sound waves become bunched up as their source approaches closer and closer to you. If the waves are bunched up and have a shorter wavelength, they have a higher pitch. Similarly, when the sound source travels away from you, the waves are stretched out (figure 1.5). Longer, more stretched-out waves have a lower pitch.

The Doppler shift applies to light waves as well as to sound waves. If the source is moving very rapidly toward us, the light waves will be bunched up and have a shorter wavelength (which corresponds to the blue and violet end of the light spectrum). However, if the light source is rapidly moving

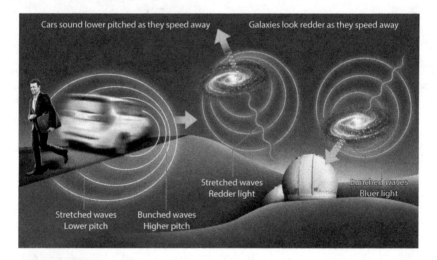

Figure 1.5 ▲

The Doppler effect occurs whenever there is motion between a source of waves and the observer. For example, when a moving car honks its horn, the sound waves appear to rise in pitch as the car approaches you and drops in pitch as the car moves away. The higher pitch is due to the compression of the sound waves as the horn approaches (bunching up the waves and shortening the wavelength). When the sound source is moving away from you, the waves are stretched out and have a longer wavelength, and the pitch drops. The same effect applies to the light waves from distant stars. If the stars were approaching us, their wavelengths would bunch up and shorten, shifting to the blue-violet end of the spectrum. However, all the stars and galaxies are shifted to the red end of the spectrum, showing that they are moving away from us. (By permission of Oxford University Press)

away from us, the light waves will be stretched out longer (which corresponds to the red end of the spectrum).

Slipher's first observations in 1912, and then Hubble and Humason's careful catalog of more than 46 galaxies and many stars, showed that almost all of the observed galaxies were red-shifted; there were almost no blue-shifted objects that might be moving toward us. More important, Hubble and Humason found that the objects farthest from us had the greatest red shifts and must be moving away from us the fastest. Hubble and Humason's work suggested that the universe was expanding. It's analogous to making a loaf of raisin bread. When you start with the ball of dough, the raisins are all packed close together. But as the ball of dough expands, each raisin moves apart from every other raisin, and those raisins on the outer part of the ball of dough move the fastest.

The universe is expanding. This is a staggering thought, and at first most astronomers were not able to accept it. However, Hubble and Humason's data were solid, and as time went on and more and more objects were analyzed, they all turned out to be red-shifted. In 1927, Belgian astronomer and Catholic priest Georges Lemaître postulated a model in which the universe had expanded from a single point in the far distant past. Most astronomers did not like the idea that the universe had a beginning; they thought it was in a "steady state" of expansion, with new matter created at the center all the time. One of these steady state advocates, Fred Hoyle, coined the term "Big Bang" to mock Lemaître's model, and that name has stuck ever since.

The controversy of Big Bang versus steady state continued for about 30 years, until the late 1950s, without any clear consensus. Then a crucial discovery was made—not by astronomers but purely by accident by two physicists and engineers, Arno Penzias and Robert W. Wilson (figure 1.6). In 1964, Penzias and Wilson were employed by Bell Labs, the original research division of AT&T/Bell Telephone, which was responsible for improving the technology of communication for "Ma Bell." They were working on improving the first antennas for receiving and transmitting signals by microwave, primarily to enable communication with NASA's Project Echo (the first attempt to use satellites for global communication), and later with the Telstar satellite. As the chief scientists and engineers on the project, their main job was to get the "bugs" out of the device and improve its efficiency. They found and eliminated many sources of "noise" from the antenna, but then they found a source of "background hiss" that was

Figure 1.6 ▲
Robert W. Wilson (*left*) and Arno Penzias (*right*) in front of the horn of their microwave antenna, which picked up evidence of the cosmic background radiation from the Big Bang. (Courtesy of Wikimedia Commons)

100 times stronger than they expected. It was detected day and night and was evenly spread across the sky (so it was not coming from a single point source on Earth or in space). It was clearly from outside our own galaxy, and they could not explain it.

Luckily just 37 miles away in Princeton, New Jersey, physicists Robert Dicke, Jim Peebles, and David Wilkinson were working on a related problem. In the 1940s, George Gamow and Ralph Alpher had predicted the existence of background "noise" left over from the Big Bang, when everything exploded with a big blast of radiation. The Princeton scientists were just beginning their experiments to detect this noise when a friend told Penzias that he'd seen a preprint of a paper by the Princeton group that predicted the exact same background noise. The two groups got in touch, and Penzias and Wilson showed them what they had found. Lo and behold, the two Bell Lab scientists had accidentally discovered proof that the Big Bang had actually happened. For this discovery, Penzias and Wilson eventually received the 1978 Nobel Prize for Physics—and this discovery had been made entirely by accident!

These stories are classic examples of how "pure" scientific research leads to amazing discoveries. Sometimes discoveries are made by people looking for an answer to a specific problem. But more often than not, scientists and engineers make important breakthroughs doing pure research—research for its own sake. Most of the best science is done by gathering a broad range of data on a particular topic without knowing what we might find. Politicians and many other people often scoff at research that does not have a definite goal in mind and try to withhold funding from these projects. But pure research is how nearly all the greatest discoveries of science are made, and science would come to an end without it—and so would all the scientific breakthroughs and life-saving discoveries that benefit us all.

Since this discovery, the Big Bang model has undergone many modifications as physicists use the properties of matter and the equations of physics to figure out how it all happened. The most recent methods date the Big Bang at about 13.8 billion years ago. At the very beginning, the universe was in a "singularity"—an infinitely small high-energy region with an infinite density. Ten milliseconds after singularity, the universe was filled with high-energy particles at temperatures over 1 trillion K that were expanding rapidly in all directions. It was so hot that only radiation, without matter, existed; space and time did not yet have the meaning we give them today, but they were infinitely warped around this extremely dense region. Over the next few seconds, the universe cooled enough to form subatomic particles, and matter, in the form of atoms, appeared in about 380,000 years. Expansion continued over the next 12 billion years, and random clumps of matter began to coalesce to form stars and galaxies and quasars. Some of these stars have already burned out and exploded, producing the heavier elements such as oxygen, silicon, carbon, iron, and so on that make up most of the matter in the solar system. In that sense, we are all stardust.

Astronomy has taught us that the universe is enormous and immense beyond our comprehension and that humans are just a tiny part of it. Our cosmic arrogance, inherited from thousands years of culture and mythology, has been dealt the death blow. The idea that we are at the center of the universe and that the universe was created for us has crumbled under the relentless effort to discover what the universe is really like—not what we would like it to be. As Carl Sagan put it in *Cosmos*:

> For as long as there [have] been humans we have searched for our place in the cosmos. Where are we? Who are we? We find that we live on an insignificant planet of a hum-drum star lost in a galaxy tucked away in some forgotten corner of a universe in which there are far more galaxies than people. This perspective is a courageous continuation of our penchant for constructing and testing mental models of the skies; the Sun as a red-hot stone, the stars as a celestial flame, the Galaxy as the backbone of night.

FOR FURTHER READING

Bartusiak, Marcia. *The Day We Found the Universe*. New York: Pantheon, 2009.

Bembenek, Scott. *The Cosmic Machine: The Science That Runs Our Universe and the Story Behind It*. New York: Zoari Press, 2017.

Brockman, John, ed. *The Universe: Leading Scientists Explore the Origin, Mysteries, and Future of the Cosmos*. New York: Harper Perennial, 2014.

Carroll, Sean. *The Big Picture: On the Origins of Life, Meaning, and the Universe Itself*. New York: Dutton, 2016.

An Illustrated Guide to the Cosmos and All We Know About It. New York: Chartwell Books, 2017.

Hazen, Robert. *The Story of Earth: The First 4.5 Billion Years from Stardust to Living Planet*. New York: Penguin, 2013.

Krauss, Lawrence. *The Greatest Story Ever Told—So Far: Why Are We Here?* New York: Atria Books, 2017.

——. *A Universe from Nothing: Why There Is Something Rather Than Nothing*. New York: Atria Books, 2012.

Natarajan, Priyamavada. *Mapping the Heavens: The Radical Scientific Ideas That Reveal the Cosmos*. New Haven, Conn.: Yale University Press, 2016.

Perlov, Delia, and Alex Velenkin. *Cosmology for the Curious*. Berlin: Springer, 2017.

Ryden, Barbara. *Introduction to Cosmology*. Cambridge: Cambridge University Press, 2017.

Sagan, Carl. *Cosmos*. New York: Ballantine, 2013.

Saraceno, Pablo. *Beyond the Stars: Our Origins and the Search for Life in the Universe*. New York: World Scientific, 2012.

Silk, Joseph. *The Big Bang*. 3rd ed. New York: W. H. Freeman, 2001.

Singh, Simon. *Big Bang: The Origin of the Universe*. New York: Harper Perennial, 2005.

Sobel, Dava. *The Glass Universe: How the Ladies of the Harvard Observatory Took the Measure of the Stars*. New York: Viking, 2016.

Tyson, Neil deGrasse, and David Goldsmith. *Origins: Fourteen Billion Years of Cosmic Evolution*. New York: Norton, 2004.

THE ABYSS OF TIME

The result, therefore, of our present enquiry is that we find no vestige of a beginning-no prospect of an end.

—JAMES HUTTON, *THEORY OF THE EARTH* (1788)

The mind seemed to grow giddy by looking so far into the abyss of time.

—JOHN PLAYFAIR (1805)

[The concept of geologic time] makes you schizophrenic. The two time scales-the one human and emotional, the other geologic-are so disparate. But a sense of geologic time is the important thing to get across to the non-geologist: the slow rate of geologic processes-centimeters per year-with huge effects if continued for enough years. A million years is a small number on the geologic time scale, while human experience is truly fleeting-all human experience, from its beginning, not just one lifetime. Only occasionally do the two time scales coincide.

—ELDRIDGE MOORES, IN JOHN MCPHEE'S *ASSEMBLING CALIFORNIA* (1993)

Since the beginning of recorded history, people have had different notions of when things happened in the distant past. In India and some parts of southern and eastern Asia, many cultures thought of time as being eternal and cyclic. The earth and life have no beginning or end; they are part of an unending cycle. Other cultures had creation myths, explaining how the universe began at some unique point of time in the past. In some Japanese creation myths, a jumbled mass of elements appeared in the shape of an egg, and later in the story Izanami gave birth to the gods. In Greek myths, in the beginning the bird Nyx laid an egg that hatched into Eros, the god of love, and the shell pieces became Gaia and Uranus. In Iroquois legend, Sky

Woman fell from a floating island in the sky because she was pregnant and her husband pushed her out. After she landed, she gave birth to the physical world. The Australian aborigines believed in a Sun-Mother who created all the animals, plants, and bodies of water at the suggestion of the Father of All Spirits.

Once the idea of an original beginning or creation event that founded the universe became part of Western culture, the next question asked was "How long ago did it occur?" Most cultures could not imagine that the universe was more than a few thousand years old. They viewed the entire universe with Earth at the center and the stars fixed to a great "celestial dome" (see figure 1.1). They thought the planets that wandered across the sky (*planetos* means "wanderers" in Greek) and the sun and the moon were carried on great wheels around Earth, as they appeared to "move" against the background of the "fixed" stars.

They thought that Earth itself was created exactly as we see it today; it was perfect and had not changed since it was formed. Any evidence of change—eroding or crumbling away—was explained as being the result of Adam's sin. Until the early 1600s, nearly all people in the Western world thought of the earth as the center of the universe, only a few thousand years old, and unchanged since its formation except for the decay due to the fall of Adam. The thinking of prominent natural historian John Woodward was typical of his time. In 1695 he wrote: "The terraqueous globe is to this day nearly in the same condition that the Universal Deluge left it; being also like to continue so till the time of its final ruin and dissolution, preserved to the same End for which 'twas first formed." A few decades earlier, in 1654, Archbishop James Ussher, the Anglican archbishop of Armagh, Ireland (which was mostly Catholic then, with few Anglicans), used the ages of the Patriarchs in the Bible to calculate that creation happened on October 23, 4004 BCE. Another scholar, John Lightfoot, placed the time as 9 A.M. This was the background for a revolution in thinking known as *geologic time*.

How has our concept of time, and our sense of the age of the universe, changed since then? Most people think of time in days or hours or minutes; if we wish to look back, we may think of time in decades or a century at most. Most humans live no more than 70 or 80 years, and only a few live to a century. Human events of more than two thousand years ago are considered "ancient," and we have a hard time comprehending the world of the Middle Ages, let alone the lives of the ancient Egyptians, Greeks, or Romans. Events of more than 5,000 years ago seem incomprehensible to us.

Contrast this concept of time with the way scientists see the world. We routinely deal in millions or even billions of years. When looking at events millions of years ago, a few hundred thousands years either way is considered unimportant. In most cases, we can't resolve the time of events that happened that far in the past, thousands or hundreds of thousands of years ago. Scientists (especially astronomers and geologists) deal in immense amounts of time, so huge that writer John McPhee called it "deep time." The epigraph (from one of McPhee's books) at the beginning of this chapter captures the essence of the problem of comprehending geologic time.

Humans are accustomed to thinking only about the short-term and the immediate future, and we have a hard time even grasping the concept of millions of years. Perhaps an analogy will help. One of the most famous is to squeeze all 4.6 billion years of geologic time into the length of an American football field, 100 yards or 300 feet, with 1 inch representing 1.4 million years. On this scale, 1 yard (3 feet) is 50 million years, and 50 yards (half the field) is 2.3 billion years. When you examine the major events of geologic history on this scale, you will be amazed at how much time went by before visible fossils (Precambrian time) appeared and how short the interval of time is for all the events that are familiar to us. If the kick returner caught the ball on the goal line, he would have run 88 yards across the field through all of Precambrian time before the first multicellular animals, such as trilobites, show up—only 12 yards from a touchdown. Just inside the 5-yard line (less than 5 yards from the goal line) is the beginning of the Age of Dinosaurs (the Mesozoic), and players would have to run to 1.5 yards from the goal line to reach the end of the Age of Dinosaurs—when they all vanished (except for their bird descendants). The entire Age of Mammals occurs in the final 1.5 yards, and the first members of the human lineage arrive only 8.3 inches from the goal line. The Ice Ages begin only 3.6 inches from the goal line. The first member of our own species, *Homo sapiens*, appears about 0.3 of an inch before the goal line. All of the last 5,000 years of human civilization is only 0.08 inches thick—narrower than a blade of grass. If the chalk stripe that marks the goal line is just a tiny bit too wide, it wipes out all of human history.

Here is another analogy. Let's squeeze the entire 4.6 billion years of Earth's history down into a single calendar year—365 days. When you divide 4.6 billion years into 365 slices, each day represents 12.3 million years. Each hour is equivalent to about half a million years (513,660 years, to be precise), and each minute is 8,561 years long. If Earth begins on New Year's

Day, then the first simple bacteria do not appear until February 21. The months roll by, with no life more complicated than single-celled organisms, until we reach October 25, when the first multicellular animals (such as trilobites and sponges) appear. Geologists call this the Cambrian Period. By November 28, we have reached the Devonian Period, when the seas were full of huge predatory fish and the first amphibians crawled out on land, cloaked by the first true forests.

By December 7 (Pearl Harbor Day), we have only reached the Permian Period, about 250 million years ago, when Earth had a single supercontinent called Pangea that stretched from pole to pole and a single ocean covered almost three-quarters of the globe. The land was dominated by huge amphibians the size of crocodiles, a variety of primitive reptiles, and huge fin-backed relatives of mammals. By December 15, we reach the Jurassic Period, a name familiar from a number of hit movies, when huge dinosaurs roamed the planet and the earliest mammals, lizards, and birds arrived. The Age of Dinosaurs ends on Christmas Day, when catastrophic events wiped out the huge dinosaurs as well as many important groups in the oceans, such as the marine reptiles. The entire past 66 million years of the Age of Mammals can be squeezed into the final week between Christmas and New Year's. The earliest human relatives do not appear until 7 hours before midnight on New Year's Eve, and the earliest members of our genus (*Homo*) are found only 1 hour before midnight. All of human civilization flashes by in the last minute of the countdown to New Year's Eve. If you start celebrating a few seconds too early, all of human history is drowned out.

Putting it this way is very humbling for humans, and for our exaggerated sense of self-importance. We are afterthoughts, very late arrivals on the stage of Earth's history, and we haven't even been around as long as most species in the fossil record. The human lineage can be traced back for only about 7 million years, whereas dinosaurs dominated the planet for more than 130 million years. Think about that the next time you hear someone use the word "dinosaur" to indicate something that is old and obsolete. We should be lucky to last as long as most species on the planet. As legendary author Mark Twain put it (with his characteristic caustic wit and sarcasm) in "Was the World Made for Man?":

If the Eiffel tower were now representing the world's age, the skin of paint on the pinnacle-knob at its summit would represent man's share of that age;

and anybody would perceive that that skin was what the tower was built for. I reckon they would. I dunno.

How did we discover the immense age of Earth and of the universe? The realization of the immensity of geologic time is just over 200 years old, and it did not come easily. Although some ancient Greeks and Romans thought Earth was really old, our modern insight about the age of Earth doesn't begin until the late 1700s, when the Age of Enlightenment (sometimes referred to as the Age of Reason) swept across Europe. It was an age of scientific discoveries, and scholarly research was less constrained by the influence and restrictions of the Catholic Church or the nobility. Enlightenment thinkers such as Voltaire, Montesquieu, Jean-Jacques Rousseau, and Denis Diderot in France, and philosophers such as George Berkeley, Jeremy Bentham, John Locke, and scientist Isaac Newton in England, were very influential. They focused on evidence and reason and the scientific method over supernaturalism and myths that had been handed down for centuries.

Surprisingly, one of the hotbeds of intellectual ferment was Edinburgh, Scotland. Even though it was not the major capital of a large country. Scotland had one of the highest literacy rates in the world at the time because the Presbyterian Church that ruled parts of Scotland believed everyone should be able to read and interpret the Bible for themselves and not have to depend on clergy to read it for them. The churches set up public schools and tried to make sure every Scot, no matter how lowly, was able to read and write. Scots had a thirst for knowledge, and great libraries and many publishing companies churned out books and newspapers. Thanks to the relatively weak influence of the many different kinds of churches in Scotland, there was no oppression by clergy as there was in England or France or much of Europe. Consequently, in the late 1700s, Edinburgh was the capital of the remarkable "Scottish Enlightenment." Brilliant and original thinkers such as the historian and philosopher David Hume (founder of modern skepticism), Adam Smith (who wrote *The Wealth of Nations*, the first great book explaining capitalism), chemist Joseph Black, and inventor James Watt (who built the first practical steam engine that launched the Industrial Revolution) all lived in the same part of Edinburgh. Most of them were drinking buddies in the social clubs, where they debated ideas and argued about science, philosophy, religion, government, and many other topics without fear of persecution. These people influenced America's founding fathers, such as Benjamin Franklin and Thomas Jefferson, who visited England, France, and Scotland and met many of the leading thinkers. Much of the Declaration of Independence and the U.S. Constitution comes directly from the thinking of John Locke and the French philosophers.

One of the geniuses of the Scottish Enlightenment was a gentleman and landowner named James Hutton (figure 2.1). He was born in Edinburgh

Figure 2.1 ▲
James Hutton in 1776, as painted by Henry Raeburn. (Courtesy of Wikimedia Commons)

on June 3, 1726, and died there on March 26, 1797. Hutton was famous in his time as a chemist and a naturalist, and he is now considered to be the "father of modern geology." Although he was trained in the law and also in medicine, he was more interested in chemistry and natural history, and he pursued those hobbies with a very independent way of thinking.

As a landowner who had inherited several large farms around Scotland, Hutton used his training in chemistry to decide how to fertilize his fields, and he traveled widely seeking new methods to improve farming practices. Meanwhile, his curiosity led him to make many observations about the slow process of weathering, how soils form, and how sediments are slowly washed out to sea and then pile up layer by layer. Eventually, he leased his properties to tenant-farmers and returned to Edinburgh to mingle with other great thinkers such as Adam Smith and Joseph Black, two of his closest friends. He traveled widely around Scotland, adding to his storehouse of observations and seeking answers to his questions about how the earth worked. He published his ideas in 1788 in a scientific paper titled "Theory of the Earth; or an Investigation of the Laws Observable in the Composition, Dissolution, and Restoration of Land Upon the Globe," and then again in a book, *Theory of the Earth*, published in 1795.

Hutton realized that everything he had witnessed demonstrated that Earth processes operated very slowly and gradually. Thick soils took years to form; layers of sediment took centuries to build up on the bottom of a lake. He visited Hadrian's Wall (figure 2.2), built by the Romans across Scotland more than 1,500 years earlier, and saw no signs that the stones had changed or even weathered much in all those centuries. In addition, he applied the Enlightenment philosophy of naturalism to geology and reasoned that the natural processes we see operating today—slow weathering, erosion, transportation of sediments—must have operated the same way in the geologic past. Ancient rocks can be explained in terms of observable processes, and those processes now at work on and within the Earth have operated with slow, steady uniformity over immensely long periods of time. This came to be known as uniformitarianism—the uniformity of natural processes through time. One of Hutton's followers, Archibald Geike, summarized the concept this way: "The present is the key to the past." We must use our understanding of present-day natural laws and processes to understand those that happened in the past.

Figure 2.2 ▲

James Hutton was familiar with the Roman barrier built across the southern border of Scotland, known as Hadrian's Wall, which was constructed about 122 CE. He was impressed that it showed few signs of weathering in the 1,500 years since it was built, and from this he reasoned that the rocks on the earth's surface must also weather and erode very slowly over hundreds to thousands of years. (Photograph by the author)

Hutton's thoughts were especially stimulated when he saw outcrops of what are known today as angular unconformities. The rocks on the bottom of this formation are tilted up on their side at an angle, then eroded off the top, and then younger rocks were deposited on that old erosional surface (figure 2.3). Hutton reasoned that the lower layers had once been laid down horizontally in the bottom of a river or the ocean. They had then been turned from soft sand and mud into hard sandstone and mudstone, a process that takes millions of years. Some time later these layers had been tilted

Figure 2.3 ▲

The famous angular unconformity at Siccar Point, Berwickshire, on the east coast of Scotland. Hutton first saw this outcrop on his last major field excursion, and it clinched his convictions that Earth was really old. Hutton realized that the bottom sequence of rocks must have originally formed as a thick stack of sands and muds on the bottom of the ocean, which were compressed into sandstones and shales. Then they were gradually tilted into a vertical orientation. Some time later, these rocks were uplifted into a mountain range, where they were eroded off, creating the erosional surface that cuts off the edges of the tilted layers. Then the entire erosional surface sank down again and was buried by younger sands, now turned into sandstones. Hutton realized that each of these processes takes thousands of years or longer, and together they indicate that the entire unconformity must represent immense amounts of time. (Courtesy of Wikimedia Commons)

on their side by immeasurably strong forces, then uplifted into the air as mountains and eroded away, as we see happening in our mountains today. Finally, those same mountains must have eroded down or sunk down, and they were then covered by an even younger layer of sediments. Knowing how slow the modern rates of weathering and erosion were, and how long it must take to deposit thousands of layers of sediment, Hutton realized that an angular unconformity must represent thousands to millions of years of time, not the mere 6,000 years that religious scholars believed. As Hutton put it, he saw "no vestige of a beginning, no prospect of an end." The earth was unimaginably old and operated on a timescale that humans could barely comprehend. Hutton claimed that the totality of these geologic processes could fully explain the current landforms all over the world, and no biblical explanations were necessary in this regard. Finally, he stated that the processes of erosion, deposition, sedimentation, and uplifting were cyclical and must have been repeated many times in Earth's history. Given the enormous spans of time needed for such cycles, Hutton asserted that the age of the earth must be inconceivably great.

For a full century after Hutton's work, no one could determine how old Earth really was. Geologists added up the maximum thicknesses of rocks representing each period on the geologic timescale, using typical rates of sediment accumulation, and tried to calculate the minimum amount of time since the Cambrian Period. Typically, their estimates were that Earth was about 100 million years old—off by a factor of 5 at least. Why were they so far off? There were lots of erosional gaps, or unconformities, in the rock record where no rock had been deposited to represent enormous amounts of time. Irish physicist John Joly estimated the age of Earth by calculating how long the world's oceans would need to accumulate their volume of salt (about 3.5 percent of normal seawater), given normal rates at which salt erodes from the land and is deposited in the ocean. His estimate was also about 80–100 million years, off by a factor of 50. As before, there were faulty assumptions in this estimate. In this case, we know that the concentration of salt in the ocean does not change much through time but stays in a stable equilibrium. Excess salt is withdrawn from the ocean when big evaporite deposits are formed, locking this salt in the earth's crust.

The most famous estimate was made by the legendary physicist William Thomson, better known by his title Lord Kelvin. He assumed that Earth started as a molten mass about the same temperature as the sun, and he calculated its age by measuring the rate of heat escaping from the heat flow coming from the earth's interior. His estimate was only 20 million years, way too short for most geologists or for Charles Darwin to accept. During much of the late 1800s, geologists began to bias their own estimates toward shorter and shorter times to appease Kelvin. Physics envy was just as powerful then as it is now!

What was the problem with Kelvin's estimate? Once again, he made a couple of bad assumptions: (1) that all the heat we measured coming up from the earth's interior was original heat from the earth's formation, and (2) no additional source of heat existed. In 1896, Henri Becquerel discovered radioactivity, and by 1903 Marie and Pierre Curie had found that radioactive materials like radium produced a lot of heat in the process of nuclear decay. In 1904, Cambridge University physicist Ernest Rutherford made many discoveries about radioactivity. He was getting ready to give a talk to the Royal Institution of Great Britain about his work when he realized that the 80-year-old Lord Kelvin himself was sitting in the audience. As a young scientist, it was frightening to be speaking in front of Kelvin himself and showing how he was wrong. As Rutherford wrote later:

> I came into the room which was half-dark and presently spotted Lord Kelvin in the audience, and realised that I was in for trouble at the last part of my speech dealing with the age of the Earth, where my views conflicted with his. . . . To my relief, Kelvin fell fast asleep, but as I came to the important point, I saw the old bird sit up, open an eye and cock a baleful glance at me. Then a sudden inspiration came, and I said Lord Kelvin had limited the age of the Earth, provided no new source [of heat] was discovered. That prophetic utterance referred to what we are now considering tonight, radium! Behold! The old boy beamed upon me.

Kelvin's estimate had been based on the faulty assumption that there were no other sources of heat beyond Earth's original heat when it cooled from a molten mass and that it would cool in no more than 20 million years. But radioactivity provides additional heat. In fact, radioactivity provides so much heat that it is now the only source of heat we measure coming from the earth's interior. The original heat from the cooling of Earth Kelvin thought he was measuring dissipated billions of years ago, maybe even during the 20 million years since Earth first formed 4.6 billion years ago.

The discovery of radioactivity not only provided a previously unknown source of heat to explain why Kelvin was wrong but also provided something else—the method to find the true age of Earth. Radioactive decay acts as a sort of clock. As unstable elements such as uranium decay, they produce stable daughter atoms such as lead. To calculate the age of a rock, all you need to do is measure the amount of parent uranium and daughter lead in the sample and the rate that it takes for a parent atom to decay to a daughter atom.

The first to realize this was Yale University chemist Bertram Boltwood, who noticed that the older his samples were the more lead they contained. By 1907, he had samples that ranged from 40 million years in age to over 2.2 billion years old. Then the brilliant young British geologist Arthur Holmes took the method even further. During his Christmas holidays in December–January 1909–10, he painstakingly analyzed a series of rocks and refined the uranium-lead dating method, solving the problems that had stymied Boltwood. By 1913, Holmes had so many results that he published the groundbreaking book *The Age of the Earth* while he was still a graduate student. (He finally got his doctorate in 1917.) He had samples from Britain that dated to 1.6 billion years old, and as he continued to date rocks through his long career teaching at Durham University and eventually at the University of Edinburgh (coming the complete circle to Hutton's own alma mater), he obtained rocks of older and older ages. The oldest were up to 4.5 billion years old, very close to our present estimate. The oldest materials we have ever dated are chondritic meteorites from the original solar system before the planets formed, and they all have dates of about 4.567 billion years.

We've come a long way since the days when we thought Earth was a flat disk in the center of the universe, that the planets and sun moved around us, and that the stars were just pinpoints of light in the celestial dome of the heavens. We now know how tiny and insignificant we are on the scale of space and in the context of geologic time. It's a humbling vision, but this is what science has revealed to us. However, there is a flip side to this coin: We are the only species that has ever been able to see and understand how we got here, and how Earth and the universe were formed.

As Carl Sagan said in the TV show *Cosmos*:

The size and age of the Cosmos are beyond ordinary human understanding. Lost somewhere between immensity and eternity is our tiny planetary home.

In a cosmic perspective, most human concerns seem insignificant, even petty. And yet our species is young and curious and brave and shows much promise. In the last few millennia we have made the most astonishing and unexpected discoveries about the Cosmos and our place within it, explorations that are exhilarating to consider. They remind us that humans have evolved to wonder, that understanding is a joy, that knowledge is prerequisite to survival. I believe our future depends powerfully on how well we understand this Cosmos in which we float like a mote of dust in the morning sky.

FOR FURTHER READING

Berry, William B. N. *Growth of a Prehistoric Time Scale*. San Francisco: W. H. Freeman, 1968.

Burchfield, Joe D. *Lord Kelvin and the Age of the Earth*. New York: Science History, 1975.

Dalrymple, G. Brent. *The Age of the Earth*. Stanford, Calif.: Stanford University Press, 1991.

——. *Ancient Earth, Ancient Skies: The Age of the Earth and Its Cosmic Surroundings*. Stanford, Calif.: Stanford University Press, 2004.

Hedman, Matthew. *The Age of Everything: How Science Explores the Past*. Chicago: University of Chicago Press, 2007.

Holmes, Arthur. *The Age of the Earth*. London: Harper and Brothers, 1913.

Lewis, Cherry. *The Dating Game: One Man's Search for the Age of the Earth*. Cambridge: Cambridge University Press, 2000.

Macdougall, Doug. *Nature's Clocks: How Scientists Measure the Age of Almost Everything*. Berkeley: University of California Press, 2008.

Ogg, James G., Gabi M. Ogg, and Felix M. Gradstein. *A Concise Geologic Time Scale 2016*. Amsterdam: Elsevier, 2016.

Prothero, Donald R. *The Story of the Earth in 25 Rocks*. New York: Columbia University Press, 2018.

Prothero, Donald R., and Fred Schwab. *Sedimentary Geology: Principles of Sedimentology and Stratigraphy*. 2nd ed. New York: W. H. Freeman, 2013.

PART II

DARWIN'S EVIDENCE FOR EVOLUTION

EVOLUTION IN ACTION

Natural selection is daily and hourly scrutinising, throughout the world, the slightest variations; rejecting those that are bad, preserving and adding up all that are good; silently and insensibly working, whenever and where opportunity offers, at the improvement of each organic being in relation to its organic and inorganic conditions of life.

—CHARLES DARWIN, *ON THE ORIGIN OF SPECIES* (1859)

Biologists finally began to realize that Darwin had been too modest. Evolution by natural selection can happen rapidly enough to watch. Now the field is exploding. More than 250 people around the world are observing and documenting evolution, not only in finches and guppies, but also in aphids, flies, grayling, monkeyflowers, salmon and sticklebacks. Some workers are even documenting pairs of species—symbiotic insects and plants—that have recently found each other, and observing the pairs as they drift off into their own world together like lovers in a novel by D. H. Lawrence.

—JONATHAN WEINER, "EVOLUTION IN ACTION" (2005)

A common myth you hear among people who don't understand evolution is that it all happened in the past but is not happening today. Nothing could be further from the truth! Evolution is not just some sort of wild guess to explain events that happened long ago. Evolution is a real phenomenon that has been documented in nature hundreds of times, by dozens of biologists working in harsh field conditions year after year, painstakingly documenting what Darwin predicted. In 1994, Jonathan Weiner wrote a Pulitzer Prize–winning book, *The Beak of the Finch: A Story of Evolution in Our Time*, which describes dozens of examples of evolution in real time. More examples have accumulated since then, and David Mindell provides many contemporary examples of evolution in his 2006 book, *The Evolving World: Evolution in Everyday Life.*

Weiner's book starts with one of the most famous examples of evolution: the Galápagos finches. During his voyage around the world on the HMS *Beagle* from 1831 to 1836, the young Charles Darwin (figure 3.1) spent

Figure 3.1 ▲

Portrait of Charles Darwin in his late twenties, painted by George Richmond in the late 1830s, after Darwin had returned from his *Beagle* voyage and begun to publish his observations and embark on his career as a scientist. (Courtesy of Wikimedia Commons)

five weeks on the Galápagos Islands during September and October 1835. The ship stopped at the Galápagos, both to explore the islands and to take on fresh water and food and do minor repairs after sailing around the stormy southern tip of South America. During that time, Darwin sailed from island to island in a small boat, wandering around and making observations, trapping animals, and shooting bird specimens to bring home.

Many ships and their the sailors used to take live tortoises on board to provide fresh meat on long voyages. Darwin noticed some obvious differences between the giant tortoises on the islands, and the British vice-governor of the islands told Darwin that he could tell them apart. Tortoises from the well-watered islands had the normal dome-shaped shells (figure 3.2A), but those from drier islands had a saddle-shaped fold that peaked on the front of their shell (figure 3.2B). This allowed them to raise their necks and heads to reach higher vegetation and cactus pads in times of drought.

Darwin observed these differences at the time and also noticed that mockingbirds differed on each island. He took careful notes about the mockingbirds and where they came from, recognizing that the birds on Chatham Island looked like the ones from South America but that those on Charles Island were very different. Mockingbirds from Albemarle Island and from James Island also appeared to be different species. He saw the remarkable marine iguanas, which are able to swim in the surf among the rocks and graze on algae, and noticed their similarity to the more typical land iguanas—but marine iguanas had clearly begun to change to an entirely unique lifestyle among lizards. As Darwin's notebooks show, after leaving the Galápagos Islands and spending many long boring days at sea, he began to speculate about the idea that species were not fixed and stable but were capable of changing—this was considered impossible at the time because most people believed God had created each species, and they all remained the same through time. The seeds for his great idea were percolating in his head, but he would not fully publish them for another 24 years (although he did record the curious features about the Galápagos Islands in his 1839 book, *Voyage of the Beagle*).

What happened in the Galápagos Islands was not fully understood until science historian Frank Sulloway painstakingly documented where Darwin was each day, and what he collected. Darwin and his manservant Syms Covington shot and stuffed dozens of birds, some of which he thought were wrens, "gross-beaks," blackbirds, or finches (31 were finches

Figure 3.2 ▲

Galápagos tortoises have differently shaped shells, depending on which island they live on. (A) Those on well-watered islands have normal dome-shaped shells and do not have to reach high to find food. (B) Tortoises from drier islands have a high saddle in the front of their shell, so they can crane their neck up to reach high vegetation and cactus pads in times of drought. (Courtesy of Wikimedia Commons)

from four different islands, representing nine different kinds). After five long years at sea, Darwin returned to England on October 2, 1836, and soon began to make arrangements for the study of his *Beagle* collections. On January 4, 1837, Darwin donated his specimens to the Zoological Society of London (they are now at the Natural History Museum in London). He had arranged for ornithologist John Gould of the British Museum to study and publish on his stuffed bird collection. (Darwin did this with most of his collections because he was not yet an expert in any field. For example, all of his fossils were given to anatomist Richard Owen, the foremost paleontologist in Britain at the time.) Gould laid aside all his other work so he could report to the assembled zoologists at the meeting a week later (January 10) that the "wrens," "gross-beaks," "blackbirds," and other birds Darwin had collected were all new species of finches. (Darwin was not there at the time, but up in Cambridge.) As Gould reported, they were "a series of ground Finches which are so peculiar [as to form] an entirely new group, containing 12 species." The report was so remarkable that it made newspapers at the time.

Darwin finally met with Gould in Cambridge in March 1837 and received a full report of the months of work that he had done. Not only were the mockingbirds from different islands members of different species, but Gould finally convinced Darwin that most of the misidentified "wrens," "gross-beaks," and "blackbirds" were actually finches that had been modified so they resembled blackbirds, wrens, and grosbeaks from other parts of the world (figure 3.3). Altogether, Gould told him that 25 of the 26 land birds were new and distinct forms not found anywhere else. As historian Frank Sulloway documented, it was only then that Darwin realized he'd been careless in his collecting and had not recorded the exact island from which each specimen had come. He had been in such a hurry, making amazing observations and collecting almost everything he could, that he had not taken time to record where he had gotten each specimen. He assumed that the birds he shot from the first island he saw were the same as the ones on the second because the islands were so close together and had very similar climate and vegetation, so he had mixed them in the same bag.

Darwin consulted with ship captain Robert Fitzroy and his steward Harry Fuller, as well as Syms Covington, who had made their own collections and kept more careful notes. Eventually Darwin was able to reconstruct the

Figure 3.3 ▲

Although Darwin didn't realize this while he was in the Galápagos, the majority of the birds there are all finches, which evolved from a generalized finch ancestor blown over from South America into birds with a wide variety of bills for nutcracking, probing for insects, picking up tiny seeds, and many other tasks performed by different families of birds on the mainland. (Modified from David Lack, *Darwin's Finches* [1947; repr. Cambridge: Cambridge University Press, 1983])

location of most of his bird specimens, and sure enough, the different species of finches were all from different islands, confirming the idea that each island had its own distinct species.

In 1836, Darwin didn't try to publicize the next obvious conclusion—that the finches had transmuted from a common ancestor blown over from the mainland into different shapes, taking on the roles of grosbeaks, wrens, and other mainland birds that did not occur on the Galápagos Islands. However, by the time of his 1839 book, *Voyage of the Beagle*, Darwin was ready to go further. He wrote:

> The remaining land-birds form a most singular group of finches, related to each other in the structure of their beaks, short tails, a form of body and plumage. . . . There are thirteen species, which Mr. Gould has divided into four subgroups. All these species are peculiar to this archipelago; and so is the whole group, with the exception of one species of the sub-group Cactornis, lately brought from Bow Island, in the Low Archipelago. . . . Seeing this gradation and diversity of structure in one small, intimately related group of birds, one might really fancy that from an original paucity of birds in this archipelago, one species had been taken and modified for different ends.

From something he almost completely missed while collecting, Darwin made the Galápagos finches into one of his best examples of creatures that must have changed from a common ancestor not very long ago. It wasn't watching evolution in action in real time, but it was suggestive of how evolution must work.

The next group to do research on the Galápagos finches was a 1905–06 expedition by the California Academy of Sciences in San Francisco, which made a much larger collection over several months. These were eventually studied by famous Oxford University ornithologist David Lack in 1938 and 1939. His 1947 book, *Darwin's Finches*, provided an updated and more detailed account of the birds, incorporating many of the details of evolutionary biology during that period. Lack spent three months on the islands documenting their behavior, ecology, and physical characteristics, filling in a lot of details Darwin had missed. However, much of Lack's research was based on stuffed bird skins in museum drawers, and he had a total of only three months of field observation, which was not sufficient to watch the birds change over the course of many years.

That job fell to Peter and Rosemary Grant of Princeton University, a husband and wife team of ornithologists who committed to doing the long-term, detailed study that might catch evolution in action. From 1973 to 2012, they spent six months in the field, mostly camped on the island of Daphne Major, one of the most isolated and unspoiled of the islands in the archipelago. There they tagged, measured, and photographed every finch on that island over the course of almost 40 years, finally retiring from the work in 2012. The Grants knew from Lack's research that the shape and size of the bill is specific to the type of foods that each finch species can eat. The ones Darwin mistook for grosbeaks have thick, robust bills, adapted for cracking large, hard seeds (see figure 3.3). Finches with smaller beaks eat smaller, softer seeds, which are more abundant in rainy years, so they can feed more rapidly that the thick-billed finches. Other finches had beaks specialized for catching insects on the wing, probing trees for grubs, and one even uses a cactus spine to "fish" for grubs in their burrows.

After many years of collecting data, the Grants began to witness some surprising events. In 1977, there was a severe drought, and in the following months the thick-billed finches were favored because only the old left-over thick-shelled hard seeds that had once been ignored were available. Within two years, these finches had evolved to have even thicker, stronger beaks. They had changed much more rapidly than anyone thought possible at the time. Meanwhile, the smaller finch species without the thick beaks died off. In 1982–83, a record El Niño year brought eight months of rain instead of the usual two months. This created a huge growth of vegetation and smaller softer seeds, which favored the smaller-beaked finches. Even when a drought hit the next year, they were still feeding on the abundant surplus.

In 1981, the Grants discovered a bird they had never seen before. It was 5 grams heavier than the other finches, so they nicknamed it "Big Bird." It had a distinctive call, glossier feathers, and could eat almost any food resource, including large and small seeds, nectar, pollen, and even cacti. Although he may have been a hybrid of the medium-beaked ground finch and the cactus finch, he lived for 13 years and generated a new population of finches that only bred among themselves, apparently creating a new species in the process. So the Grants had documented not only evolutionary change within species in response to selection but also apparently watched a new species forming. Recent research has identified the genes that control beak shape in these finches and have artificially duplicated the pattern seen in nature by adding or subtracting those genes.

The Grants were both born in 1936, and they are retired and 84 years old as I write this. But they are legends in biology for their incredibly long, detailed, and difficult study, spending almost half their lives for 40 years camped on the tiny, rugged island of Daphne Major doing enormous amounts of work to document evolution in real time. They have received almost every award possible in their field, including the 1994 Leidy Award of the Academy of Natural Sciences in Philadelphia, the 2003 Loye and Allen Miller Award, the 2005 Balzan Prize for Population Biology, the 2009 Kyoto Prize in basic sciences, the 2017 Royal Medal in Biology, and, in 2008, the highest award of all—the Darwin-Wallace Medal of the Linnaean Society of London—which is bestowed only once every 50 years. And thanks to Weiner's 1995 book, *The Beak of the Finch*, they are world famous living legends among professional biologists.

The work of the Grants filled a long-standing gap in research about evolution. Darwin had shown that life had evolved, and that natural selection was the best mechanism for *how* it had evolved, but he had no way of watching it happen in real time—he could only infer its existence from the results. In 1893, German biologist August Weismann remarked "that it is really very difficult to imagine this process of natural selection in its details; and to this day it is impossible to demonstrate it in any one point."

For most of the next century, evolutionary biology grew and expanded with the discovery of genetics and mathematical modeling of population biology, showing that natural selection was a sufficient mechanism to explain how evolution occurred. But few biologists could find a way to watch it happen in real time in nature because it required years in the field and lots of slow, painstaking observations and data collection. By 1934, one geneticist commented that if "ever an idea cried and begged" for an experimental research program "surely it is this one . . . but there have been so very, very few of them." In 1960, a different geneticist commented that "the amount of observation or experiment so far carried out upon evolution in wild populations" was still "surprisingly small." He found this disturbing because "evolution is the fundamental problem of biology while observation and experiment are the fundamental tools of science." As late as 1990, an anthropologist complained in *The Encyclopedia of Evolution* that the "complaint of a half-century ago holds good: the number of experimental tests of natural selection is pitiful; the few that have been contacted still do heavy duty as exemplars."

But thanks the work of the Grants and numerous other biologists since the 1970s and 1980s, these laments are no longer true. Instead, the studies of natural selection changing existing species or new species arising have multiplied to form a major research field unto themselves. One of the first such studies was by Darwin's Cambridge classmate Benjamin Walsh. He and his wife migrated to the United States and had built and lost several farms in different states, and eventually he became a self-taught entomologist. When Darwin's book reached him, he was staggered and slowly became convinced of Darwin's conclusions.

Walsh then observed an example of a species adapting to a new resource. One species of fruit fly, known as the haw fly, laid its eggs on wild hawthorn fruits. But as farmers spread cultivated apple trees across the region, the haw fly changed its habits and adapted to feeding on apples, and it slowly spread across the landscape as a new species of apple fly. In 1867, Walsh published a paper in which he pointed out that "it attacks cultivated apples only in a certain limited region, even in the East, for . . . this new and formidable enemy of the apple is found in the Hudson River Valley, but has not yet reached New Jersey." Walsh died shortly after he published this prediction, but he was proven right. They were reported in northern New York and Vermont and New Hampshire in 1872, in Maine in 1876, and in Canada by 1907. Meanwhile, they spread south through Georgia in 1894, and west to Michigan by 1902. Eventually they spread across the entire country, reaching the West Coast only about 30 years ago. Another species evolved to eat rose hips, and sometimes pears and plums, and in northern Wisconsin another species eats sour cherries—all the while, the original haw fly still eats hawthorn fruit. Subsequently, the genetics of these different fly populations have been studied, and they are genetically distinct, showing that they have become different species in a little over a century.

Another early case of evolution is the English sparrow, which was observed by Hermon Carey Bumpus. The study was a lucky accident—a huge blizzard had hit on January 31, 1898, affecting dozens of birds, which were brought to Bumpus in his lab at Brown University in Providence, Rhode Island. In the warmth of the lab, 72 of the sparrows revived, but 64 were past saving and died. Bumpus measured several dimensions of all the birds and recorded their sex. He found that in the extreme conditions of the blizzard the surviving males were mostly shorter and lighter, selecting against the largest males (this is called directional selection). The surviving females were all of medium size, so selection had worked against both the largest and smallest members of their population (this is called stabilizing selection). This early case soon became famous and was featured in lots of biology textbooks of the twentieth century. However, it looked at only a single instance of selection and was not repeatable; other researchers were not able to confirm whether blizzards caused the population to change over time.

Wild bird populations are studied by lots of people (including amateur birdwatchers), and they have often demonstrated examples of evolution. For example, the common European house sparrow was introduced in North America in 1852 and has since spread over nearly all the populated areas of our continent, from the forests of Canada down to Costa Rica. After spreading to all of these habitats, they are rapidly diverging into populations of different body sizes, with the more northern populations being larger. This is a well-known adaptation called Bergmann's rule: Large bodies have less surface area compared to their larger volume and thus lose body heat slower than smaller, skinnier body forms that are more suitable for the tropics. There are also changes in wing length, beak shape, and other features, so these sparrows are quite different at different latitudes in the Americas, and they are often placed in different subspecies. Those that live in dry desert climates tend to be lighter tan in color, for protective camouflage. Some of them, like the wild *Bactrianus* subspecies of sparrow, have diverged genetically as well.

In some cases, transformation can be very rapid. For example, the sockeye salmon in the wild living in fast-moving rivers normally have males with slender strong bodies to swim against currents and larger females who can did deeper holes in which to lay their eggs without the river eroding them away. But in 1957, salmon colonized a beach in the Seattle area called Pleasure Point, where they were living in quiet deeper waters (figure 3.4). Within 40 years, the males evolved to have deeper, rounder bodies because they no longer had to fight strong currents, but their biggest challenge was fighting off other males for access to mates. The females also evolved smaller bodies because they do not have to dig deep holes for their eggs. Genetic study of these salmon showed that they had already diverged markedly from their ancestral species and were on their way to becoming a distinct new species.

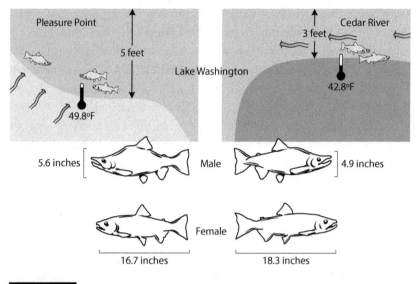

Figure 3.4 ▲
Evolution has happened in some groups over just a few decades. In the swift currents of Cedar River, Washington, the sockeye salmon introduced in the 1930s have adapted so that the males can swim in the strong currents and the females dig deeper nests in the sand to lay their eggs. But the male salmon that invaded the shallow waters of Pleasure Point in 1957 have developed rounder, deeper bodies to help fight off rival males, and the females have smaller bodies and dig shallower nests for their eggs because there are no strong currents. (Modified from Jonathan Weiner, "Evolution in Action," *Natural History* 115, no. 9 [2005]: 47–51)

A well-studied example is the three-spined stickleback fish, which can modify its body armor and number of spines depending on its habitat (figure 3.5). Sticklebacks that live in deeper ocean habitats have heavier body armor, and populations that live in lakes have lighter armor. This change occurred in less than 31 years in a pond near Bergen, Norway, and the change took only a dozen years in Loberg Lake, Alaska. Changes in their spines have also been documented. Sticklebacks that live in deeper more open water have long spines to deter predators. But long spines can be a problem for sticklebacks in streams where a dragonfly larva can grab the spines with its pincers, so stream sticklebacks tend to shrink or lose their spines. A single gene, *Pitx1*, turns the switch that regulates spine length off and on. Some studies artificially modified captive sticklebacks so they had unusual new combinations of spines. These studies found that females only

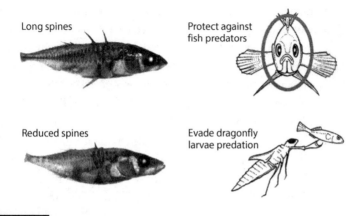

Long spines

Protect against fish predators

Reduced spines

Evade dragonfly larvae predation

Figure 3.5 ▲

The three-spined stickleback fish also show rapid evolution. Species that live in lakes or oceans have longer spines, which makes it harder for a predator to swallow them. Stickle-backs that live in shallow streams have evolved shorter spines so dragonfly larvae and other predators cannot catch them by their protruding spines. (Photographs courtesy of D. M. Kingsley and Sean Carroll)

mate with males that have new traits, so sexual selection is a driving force in their evolution of novelty. Prior to these studies, ichthyologists would read-ily assign specimens with different spine counts and different body armor to different species, but these studies show just how easy it is for one stickle-back population to transform to another species given the right conditions.

Examples like these could be multiplied endlessly. In New England, the periwinkles have dramatically changed their shell shape and thickness in less than a century, probably due to predation pressure by newly intro-duced crabs. In the Bahamas, the anole lizards (the common "chameleon" in pet shops, which are not true chameleons) changed the proportions of their hind limbs when they were introduced to new islands with different vegetation. In Hawaii, the honeycreeper evolved shorter bills as the birds switched to another source of nectar when their favorite food source, 'the native lobelloids, disappeared. In Nevada, the tiny mosquito fish that live in isolated desert water holes once connected during the last Ice Age have quickly evolved major differences in less than 20,000 years. And in Australia, the wild rabbits brought by European settlers less than a century ago have modified their body weight and ear size in response to the differ-ent conditions of the outback.

Some of the best examples of rapid evolution occur in insects, such as Walsh's study of haw bugs. Most insects have very short generation times and lay hundreds to thousands of eggs, so they can transform much more quickly than animals with long generation times and slow birth rates. Perhaps the most famous example of all is the peppered moth, *Biston betularia*, found in many biology textbooks (figure 3.6). Wild moths were naturally covered with speckles on their wings that camouflage them against the mossy bark of a tree. But this all changed during the Industrial Revolution when pollution turned the tree trunks black. A formerly rare mutant form with black wings suddenly became the most common because they were well disguised on the blackened tree bark during the height of the pollution caused by burning coal. The wild speckled variety

Figure 3.6 ▲

The peppered moth, *Biston betularia*, is normally covered with a speckled gray pattern, but it has a rare mutant form that is completely black. These black forms became dominant during the Industrial Revolution when soot caused the blackening of tree bark, making the black moths less conspicuous and therefore favored by natural selection over the speckled varieties. (Courtesy of Wikimedia Commons)

nearly disappeared because they were now conspicuous to bird predators and were selected against. In the 1960s and 1970s, antipollution laws shut down or cleaned up the power plants, and the trees returned to their normally gray mottled mossy appearance. The black moths became rare again, and the speckled moths thrived because they were again well camouflaged.

Another example of recent rapid evolution in insects is the soapberry bug, a small insect with a long "beak," or proboscis, that it uses to pierce the rind of a fruit or seed pod and suck out the nutritious interior. In the southern United States, they live on three kinds of native plants, and the bugs have relatively long beaks. But in Florida, they have learned to live on a plant that was introduced in the 1950s, the flat-podded golden rain tree, which has thinner seed walls. In the wild, their beaks are about 9 millimeters long, but they only need about 3 millimeters of beak to pierce the outside of this new food source. Consequently, the soapberry bugs in Florida have been adapting by growing shorter and shorter beaks in just the past 60 years. In another region, the bugs feed on the heartseed vine (which has a thicker shell) and have already begun to develop longer beaks since 1970 when that plant was introduced, less than 50 years ago.

Among the many examples of evolving insects, the most striking are those that have evolved resistance to pesticides, all within a few decades, causing enormous economic damage all over the world. Every modern housefly now carries genes that make it resistant not only to DDT but also to pyrethroids, dieldrin, organophosphates, and carbamates, so there are few poisons left that can suppress them. The mosquitoes that evolved resistance to DDT and other organophosphate insecticides apparently evolved in Africa during the 1960s, spread into Asia, and reached California by 1984, Italy in 1985, and France in 1986. As entomologist Martin Taylor describes it in Jonathan Weiner's book *The Beak of the Finch*:

> It always seems amazing to me that evolutionists pay so little attention to this kind of thing, and that cotton growers are having to deal with these pests in the very states whose legislatures are so hostile to the theory of evolution. Because it is the evolution itself they are struggling against in their fields every season. These people are trying to ban the teaching of evolution while their own cotton crops are failing because of evolution. How can you be a creationist farmer any more? (225)

But by far the fastest examples of evolution occur in microbes, which can reproduce and multiply in a few minutes to hours of generation time into thousands of individuals in a population. There will never be a cure for the common cold virus because the viruses that cause colds evolve new protein coats in a few months that make them unfamiliar to our immune system and thus capable of attacking us again (until our immune system eventually catches up after we suffer through a nasty cold). Every year the world's medical labs must develop new flu shots because every year several new strains of flu emerge that our immune system doesn't recognize. The flu shot contains killed versions of several strains of flu; your immune system reacts to them and is prepared so you don't get sick with the flu. But there are many new strains each year, and the flu shot can't anticipate them all, so every year even people with the flu shot get infected. This is a never-ending battle with organisms that evolve far faster than we can protect against them, just like the insects that develop resistance to all our pesticides.

So the next time you get the cold or flu, or are bitten by a mosquito even though you sprayed pesticides to keep them away, you are experiencing evolution in action. Evolution happens all around us all the time, whether we acknowledge it or not. Developing new pesticides and flu shots is essential to our survival in this race with these rapidly evolving organisms. If we don't acknowledge evolution, we will lose the race against diseases and crop loss.

FOR FURTHER READING

Levinton, Jeffrey. *Genetics, Paleontology, and Macroevolution.* 2nd ed. New York: Cambridge University Press, 2001.

Mindell, David P. *The Evolving World: Evolution in Everyday Life.* Cambridge, Mass.: Harvard University Press, 2006.

Ridley, Mark. *Evolution.* 2nd ed. Cambridge, Mass.: Blackwell, 1996.

Weiner, Jonathan. *The Beak of the Finch: A Story of Evolution in Our Own Time.* New York: Knopf, 1994.

——. "Evolution in Action." *Natural History* 115, no. 9 (2005): 47–51.

OUR COMMON BODY PLAN

We have seen that the members of the same class, independently of their habits of life, resemble each other in the general plan of their organisation. This resemblance is often expressed by the term "unity of type"; or by saying that the several parts and organs in the different species of the class are homologous. The whole subject is included under the general term of Morphology. This is one of the most interesting departments of natural history, and may almost be said to be its very soul. What can be more curious than that the hand of a man, formed for grasping, that of a mole for digging, the leg of the horse, the paddle of the porpoise, and the wing of the bat, should all be constructed on the same pattern, and should include similar bones, in the same relative positions?

—CHARLES DARWIN, *ON THE ORIGIN OF SPECIES* (1859)

The famous British philosopher and mathematician Alfred North Whitehead wrote, "the safest general characterization of the European philosophical tradition is that it consists of a series of footnotes to Plato." Of course this is an oversimplification, but it reflects the fact that Plato identified and laid the groundwork for most of the major themes in philosophy that would be further developed over the next 2,000 years. In some ways, much of modern science (especially biology) originated with Plato's famous student Aristotle, who focused not so much on philosophy as on describing and explaining nature as the ancient Greeks knew it about 350 BCE. However, unlike the eternally unsolved problems of philosophy that Plato delineated, the ideas of Aristotle have mostly faded into obscurity because modern science has proved him to be wrong more often than he was right.

Aristotle was right, however, when he discussed an important concept we now call *homology*. Although other ancient naturalists may have noticed it, Aristotle was apparently the first to describe it in writing (or, at least in ancient writings that still survive). He pointed out that the same bones in vertebrate limbs were highly modified for completely different uses (figure 4.1). For example, the structure of the tetrapod forelimb has the same basic elements: a long robust upper arm bone hinged to the shoulder

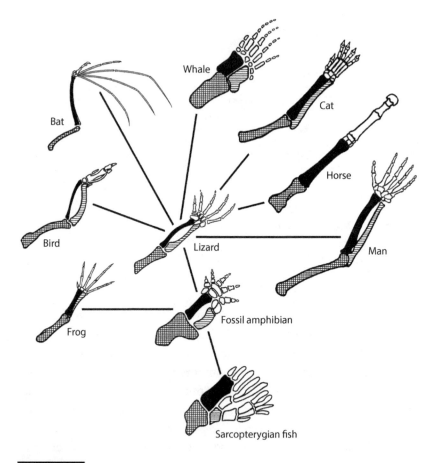

Figure 4.1 ▲

The evidence of homology. All vertebrate forelimbs are constructed on the same basic plan with the same building blocks, even though they perform vastly different functions. The basic vertebrate forelimb has been modified into a flipper in whales, a wing in bats, and a one-fingered running hand in horses, yet the basic bony structure remains the same. (Drawing by Carl Buell)

blade with the ball-and-socket joint (the humerus); a pair of long bones in the forearm (radius and ulna); and a number of bones that make up the wrist and fingers (carpals, metacarpals, and phalanges). Yet this basic structure has been modified for completely different functions in various vertebrates. In humans and other primates, the structure of these bones is relatively unmodified, but in whales the bones are modified to form a flipper; in bats, the fingers are extremely elongated to support their wing membrane; in horses, the side fingers are all lost and only the middle finger and extremely elongated middle wrist bone (metapodial) remain to form the hoof; in a mole, the wrist and fingers are very short and robust for digging; in a cat, the fingers are fused into a paw with sharp claws, and so on. In other words, the vertebrate forelimb starts with a basic body plan but is completely transformed to serve very different functions. Aristotle didn't have an explanation for this phenomenon, but he argued that it showed the basic unity of the plan of nature—it was a unified organization rather than a random hodgepodge of creatures.

Aristotle also distinguished this concept from its opposite, *analogy*. In homology, animals use the same fundamental parts for different functions; but in analogy, animals are modified to perform similar functions using very different parts of their anatomy. For example, most aquatic vertebrates have streamlined, torpedo-shaped bodies with some sort of fin to propel them (figure 4.2). One need only compare the fish-like bodies of whales and dolphins, penguins, and the extinct dolphin-like reptiles known as ichthyosaurs to see their superficial similarity. But their underlying structures are very different (particularly in their internal anatomy and reproduction), so we know that this body shape evolved independently for swimming in different groups: mammals (whales), birds (penguins), reptiles (ichthyosaurs), as well as the many lineages that are lumped into the wastebasket known as "fish."

Or consider the different ways that flying animals have built their wings (figure 4.3). Bats use their elongated fingers, but birds have fused all their finger bones into a single bone called the alula (it's the thin bony part of the chicken wing that you never eat) and support their wings with feather shafts. The flying reptiles known as pterosaurs (also called pterodactyls) have a hugely elongated fourth finger ("ring finger") to support their wing. Flying insects have built their wings out of completely different parts of the body than the arms of flying vertebrates. These structures all serve the function of flight whether in mammals, birds, reptiles, or insects, but

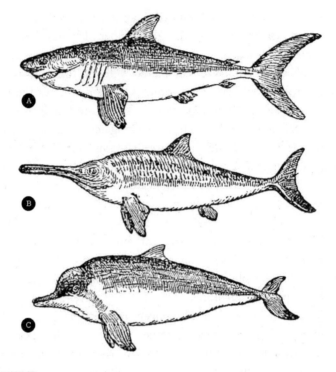

Figure 4.2 ▲

The opposite of homology is analogy (also known as convergent evolution), in which structures of unrelated organisms evolve to look and function in a similar way. The streamlined body and fin structures of aquatic animals such as fish, ichthyosaurs, and dolphins give them very similar body shapes, even though a study of their anatomy unrelated to swimming shows that one is an air-breathing mammal (figure 4.2C), another an air-breathing reptile (figure 4.2B), and the shark is a gilled vertebrate like other fish (figure 4.2A) that can extract oxygen from water directly.

they are built in totally different ways. In fact, many plant seeds, such as sycamore seeds, have wing-like seed casings that enable the falling seed to "fly" a considerable distance from the original tree—yet clearly it is built of entirely different structures than the wings of any animal.

Aristotle's demonstration of homology was discussed many times by scholars over the next two millennia. Perhaps the first really detailed account was by the French naturalist, traveler, and diplomat Pierre Belon. Like many educated Renaissance scholars, he was interested in a wide range of topics, including architecture, Egyptology, botany, ichthyology,

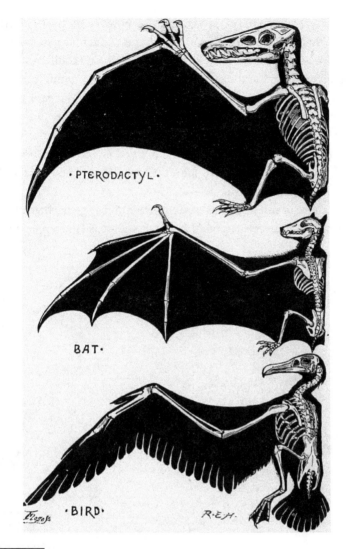

Figure 4.3 ▲

Convergent evolution is strongly demonstrated in three different groups of flying vertebrates. Bats use the highly elongated fingers of their hand to support their wing membranes,
whereas birds fingers are fused down to a single bone and they support their wings with
feather shafts. Pterosaurs evolved yet another solution—the wing is supported by a hugely
elongated fourth finger (the "ring finger").

and comparative anatomy. He wrote a 1551 book on strange marine fish (which included whales back then). In 1553 he wrote four new books, one that became the foundation of ichthyology, another on conifers, a third on funerary customs of the ancients (especially Egyptian mummification), and a fourth on "memorable things" found in Greece, Asia, Judea, Egypt, and "other strange countries." In 1555, he wrote *L'Histoire de la nature de oyseau* (History of the Nature of Birds). In it he discussed some of the first concepts of what is now called comparative anatomy, and he included a remarkable figure that was one of the first to show the homologous bones on the skeletons of birds and humans (figure 4.4).

This pattern of similarity was treated as part of the "great chain of being" for several centuries, explaining the stepwise increase in complexity from

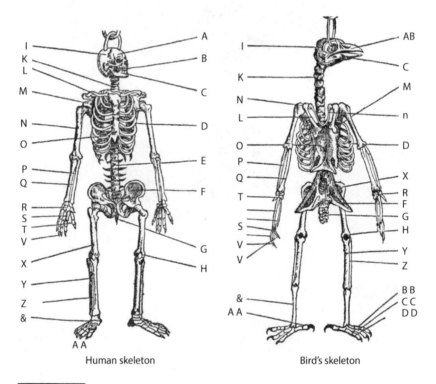

Human skeleton   Bird's skeleton

Figure 4.4 ▲

Pierre Belon's famous sketch of a human skeleton and bird skeleton drawn to the same size and in a similar pose, showing the bone-for-bone similarity of most of their bones. From *L'Histoire de la nature de oyseau* (History of the Nature of Birds, 1555).

sponges and corals to molluscs to fish to reptiles to humans—and often further up the chain of divine beings, from cherubim and seraphim to angels to archangels to God. By the 1700s, the German school of *Naturphilosophie* ("natural philosophy" or "natural history" to the English) was extolling homology as part of the great plan of unity in nature (usually with religious overtones suggesting that God had intended it that way). The great German poet, philosopher, and scholar Johann Wolfgang von Goethe noted in 1790 that even plants showed homology because the petals and sepals of flowers were just modified leaves. In 1818, the pioneering French zoologist Étienne Geoffroy Saint-Hilaire published his *theorie d'analogue*, positing that the structures shared among the vertebrates were all the same. He tried to extend this to invertebrates, and by doing so he triggered the famous debates with Baron Georges Cuvier, the greatest anatomist and paleontologist of that time, who believed there were no connections between his five great *embranchements* (branches) of nature.

The famous Estonian embryologist Karl Ernst von Baer (see chapter 5) first noticed that homologous structures, such as the vertebrate forelimb, started from embryonic structures shared in common among all vertebrates, thus tracing the roots of homology back to embryology. Finally, in 1843 the legendary British anatomist and paleontologist Richard Owen formally coined the term "homology," and it has been used in that sense ever since. His tests of whether something was homologous depended on three things: position, development, and composition. However, Owen cooked up his own idiosyncratic explanations for why homology existed: They were part of the basic "archetype," or God's blueprint for how organisms were designed.

All of these older so-called explanations for homology faltered on the same problem—they described it but really didn't explain it. In most cases, archetypes or unity of type ideas just restate the obvious: all have a common body pattern but *why* they do is not known. In many instances, the natural philosophers would say that they all have a common pattern or archetype because this was the blueprint for life in the mind of God. But that is not a testable hypothesis either; just saying "God did it" does not allow for any kind of scientific analysis or testing of that idea. This was particularly true when von Baer pointed out that homology extended down to the embryonic precursors of forelimbs.

Into this breach stepped Charles Darwin. Since his student days at the University of Edinburgh and at Cambridge University, he had been familiar with the discussion about unity of type that was advocated by the natural philosophers of his time. These things were in his mind when he returned from the five-year voyage of the *Beagle* in 1836. As his early notebooks show, the voyage and his subsequent research made him doubt that species were "fixed" and "immutable" and could not change over time. Once he had opened the door to the idea of one species transforming into another, and that this change was driven by natural selection, the unity of type could be explained. The reason for homology of the organs of so many kinds of plants and animals is that their common ancestor was built on that same body plan, and descendants had modified what they had inherited for another function.

Likewise, natural selection explained analogy. If selection pushed organisms to adapt to their local environment, then different groups from different ancestry, such as fish, whales, penguins, and ichthyosaurs, would adopt a more streamlined body form to move through water more efficiently. Animals that had become fliers would build their wings out of what they inherited from their ancestors, which is why the wings of insects are so different from those of bats, birds, and pterosaurs. Darwin himself saw how the beaks of Galápagos finches had been transformed to resemble beaks of unrelated birds on the mainland, all because they had a common function. After all, if an all-powerful Divine Designer wanted to build wings, why wouldn't he use the same, most efficient construction plan rather than jury-rig wings out of totally different structures? Today we call this development of similar structures in unrelated groups *convergent evolution*.

Since Darwin originally discussed homology in *On the Origin of Species* in 1859 (see epigraph at the beginning of the chapter), the examples have multiplied tremendously. Some of the best examples are seen in the development of wings and other appendages in the arthropods, the "jointed legged" animals that include insects, spiders, crustaceans, trilobites, and their relatives. All of these animals have a flexible body plan with a number of segments (somites), each of which can bear different appendages. They can multiply their segments as in millipedes or centipedes or modify their appendages on each segment to different functions: legs, mouthparts, antennae, pincers, and wings. The fact that each segment has basically the same structure and is repeated over and over again is known as *serial homology*. Each appendage on each segment is homologous with different kinds of appendages on other segments. The earliest known arthropods are the millipedes; later groups were developed by reducing the number of segments and changing the function of the appendages on each segment. Even within groups some remarkable examples of serially homologous organisms being modified can be found. For example, the most primitive winged insects are dragonflies and damselflies, which have two pairs of wings on their backs (figure 4.5B). But in beetles, the first set of wings has been modified into a pair of hard shell-like covers called elytra that protect the flying wings behind them by covering them up (figure 4.5A). In flies and mosquitoes and their relatives, the rear set of wings has been modified into a pair of knob-like structures called halteres that help stabilize them during flight (figure 4.5C). All of these features develop from a common embryonic stem, which would normally form a flying wing, but they change during development as the insect goes through different larval stages until it becomes an adult.

A

1 cm

Figure 4.5 ▲ ▶

Structures made from the embryonic wing precursors in insects. (*A*) The first set of wing buds become the hard protective shells, or elytra, in beetles, protecting the delicate set of flying wings developed from the second pair of wing buds. (*B*) In dragonflies, both pairs of wings are fully developed. (*C*) In flies, the first set of wings develops, but the second set is a pair of short knobs called halteres that help stabilize flight. (Courtesy of Wikimedia Commons)

(continued)

In fact, the exact pattern of embryonic development of arthropods from their primitive body plan has all been worked out, along with understanding the underlying genetic code that controls it. The first 10 somites all have numbers (table 4.1), and different arthropod groups modify the appendages on each somite differently. For example, trilobites only have their appendages on somite 1 modified into antennae, and the rest of the segments bear legs. Spiders have only six total somites: somite 1 bears the chelicerae (jaws and fangs), somite 2 becomes the pedipalps (the rest of their mouthparts), and somites 3–6 each bear a pair of legs, giving them eight legs. Centipedes produce antennae on somite 1, no appendage on somite 2, their mouthparts on somites 3–5, and the first set of legs begins on somite 7, with dozens of somites and pairs of legs following. In insects, the antennae grow on somite 1, the mouthparts on somites 3–5, and their three sets of legs on somites 6–8. Finally, crustaceans have two sets of antennae on somites 1–2, their mouthparts emerging from somites 3–5, and their five pairs of legs are on somites 6–10. Thus somites 3–5 are legs in trilobites and spiders, but mouthparts in centipedes, insects, and crustaceans.

Table 4.1 The homology of different appendages in specific groups of arthropods

Somite (body segment)	Trilobite (Trilobitomorpha)	Spider (Chelicerata)	Centipede (Myriapoda)	Insect (Hexapoda)	Shrimp (Crustacea)
	Antennae	Chelicerae (jaws and fangs)	Antennae	Antennae	1st antennae
	1st legs	Pedipalps	–	–	2nd antennae
	2nd legs	1st legs	Mandible	Mandible	Mandible
	3rd legs	2nd legs	1st maxillae	1st maxillae	1st maxillae
	4th legs	3rd legs	2nd maxillae	2nd maxillae	2nd maxillae
	5th legs	4th legs	Collum (no legs)	1st legs	1st legs
	6th legs	–	1st legs	2nd legs	2nd legs
	7th legs	–	2nd legs	3rd legs	3rd legs
	8th legs	–	3rd legs	–	4th legs
	9th legs	–	4th legs	–	5th legs

Source: Compiled from different sources by the author.

And what about the stinger on wasps and bees? The stinger originally was a long tubular injection device (like a hypodermic needle) known as the ovipositor, which many types of female insects use to lay eggs (often within another living substance such as another insect; see chapter 8). But many highly social insects such as wasps and bees no longer use it for laying eggs. Worker bees, for example, are all females, but none of them can reproduce, so their ovipositors are modified into stingers. If they sting something, it's usually a suicide mission because their stinger tears out of their abdomen, killing the bee. Male bees don't have the ability to develop a stinger at all because they don't have the genes for an ovipositor, which is a female reproductive organ.

Examples of surprising homologies are found all over biology. Right now, for example, you are hearing sounds through the three bones in your middle ear: the malleus (hammer), incus (anvil), and stapes (stirrup). They conduct sounds from the vibrations of your eardrum to the pressure plate on the end of the long coiled tube in your inner ear (cochlea), which turns these vibrations of the tiny hairs in fluid into sound in your brain. But how did this clumsy and inefficient arrangement develop? When you were an embryo, the hammer and anvil were part of the joint between the skull and jaws, and they migrated to your middle ear during your embryonic development. The anvil was originally the quadrate bone in the back of the skull, which hinged to the angular bone (= hammer) in the back of the jaw.

Why did mammals develop such a clumsy arrangement? We can trace its origin to the fossils of the earliest relatives of mammals, known as synapsids or "protomammals." These fossils were once incorrectly called "mammal-like reptiles," but they never had anything to do with reptiles and that term is now obsolete. As we trace the evolution of early synapsids into mammals, we see both the angular bone in the jaw and the quadrate bone in the skull get smaller and smaller until a new joint is developed between two different bones, the squamosal bone of the skull and the dentary bone of the jaw, which is the jaw joint you now have in your skull. These bones persisted while the other reptilian bones of the jaw were lost because reptiles usually transmit sound from their lower jaw through the jaw joint to the inner ear. Once the old reptilian quadrate-articular jaw joint was no longer functioning as a jaw hinge, it reverted to its other function, hearing (figure 4.6).

Other examples of homologies abound in all animals, including ourselves. For example, our reproductive organs (testes and penis in males,

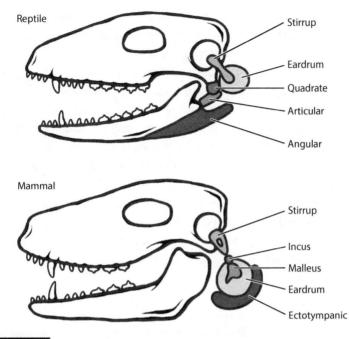

Figure 4.6 ▲

The ear region in synapsids ("mammal-like reptiles") and mammals undergoes a dramatic transformation, as the articular bone of the lower jaw hinge and the quadrate bone of the jaw hinge in the skull shift to the middle ear and become the incus (anvil) and the malleus (hammer). This same transformation can be seen not only in fossils but also during the embryology of a mammal. When you were an embryo, your middle ear bones started out in your jaw.

ovaries and clitoris in females) are homologous and perform many of the same functions. In early human embryos, our reproductive organs remain undifferentiated until the appropriate hormones are released to trigger development of the reproductive organs. If you have a "Y" chromosome, testosterone is sent to your embryonic reproductive organs to start their development into male organs. If you lack the "Y" chromosome, the normal default setting of our embryos is followed and female reproductive organs develop. Occasionally, problems occur with the hormonal signaling of the embryonic reproductive organs, and humans are born as hermaphrodites with both sets of organs partially or completely developed. But keep this in mind—we humans are all females unless we get testosterone at the right time to turn us into males!

Many more examples of surprising homologous organs could be described, but the big picture is the one to remember: Homologies are the clear sign of evolution from a common ancestor, using a common embryonic pathway. Homologous organs could not be explained by the pre-Darwinian idea of unity of type or archetypes. They clearly show evidence of how evolution modifies what the embryo already has in its tool kit to produce a different structure. It's not efficiently designed or well-engineered, but this is consistent with the idea that nature tinkers with what it already has in the way of ancestral states to produce new structures. In the 161 years since Darwin's book, we have discovered more examples of this clumsy, inefficient process, and we have learned the genetic and embryonic explanation for how and why it happens. We have come a long way since Darwin's time, and homology is an even stronger line of evidence for evolution now than it was then.

FOR FURTHER READING

Carroll, Sean. *Endless Forms Most Beautiful: The New Science of Evo Devo*. New York: Norton, 2005.

——. *The Making of the Fittest: DNA and the Ultimate Forensic Record of Evolution*. New York: Norton, 2006.

Fitch, Walter M. "Homology: A Personal View on Some of the Problems." *Trends in Genetics* 16, no. 5 (2000): 227–231.

Hall, Brian. *Homology*. Novartis Foundation Symposium, 2008.

Mindell, David P., and Axel Meyer. "Homology Evolving." *Trends in Ecology and Evolution* 16, no. 8 (2001): 434–440.

Panchen, Alec L. "Homology—History of a Concept." In *Homology*, ed. Gregory R. Bock and Gail Cardew, 5–18. Chichester, UK: Wiley, 1999.

Scotland, Robert W. "Deep Homology: A View from Systematics." *BioEssays* 32, no. 5 (2010): 438–449.

Wagner, Günter. *Homology, Genes, and Evolutionary Innovation*. Princeton, N.J.: Princeton University Press, 2014.

Zakany, Jozsef, and Denis Duboule. "The Role of Hox Genes During Vertebrate Limb Development." *Current Opinion in Genetics & Development* 17, no. 4 (2007): 359–366.

ONTOGENY RECAPITULATES PHYLOGENY

It has already been casually remarked that certain organs in the individual, which when mature become widely different and serve for different purposes, are in the embryo exactly alike. The embryos, also, of distinct animals within the same class are often strikingly similar: a better proof of this cannot be given, than a circumstance mentioned by Agassiz, namely, that having forgotten to ticket the embryo of some vertebrate animal, he cannot now tell whether it be that of a mammal, bird, or reptile. The vermiform larvæ of moths, flies, beetles, &c., resemble each other much more closely than do the mature insects; but in the case of larvæ, the embryos are active, and have been adapted for special lines of life.

—CHARLES DARWIN, *ON THE ORIGIN OF SPECIES* (1859)

Before the invention of the microscope, the true nature of animal reproduction was a complete mystery. Without the magnification of strong lenses, no one could see the sperm or egg of humans. We knew that sex led to offspring—but not how it worked. With the naked eye, we could only see how babies grew up once they were well along in their development. The ancient Greek philosopher Pythagoras (famous for the Pythagorean theorem) was one of the earliest recorded thinkers on this topic. He originated the idea of "spermism," that is, all the essential characteristics of humans were carried in the sperm and the mother only contributed the material support for the developing embryo. This idea was popularized by Aristotle and accepted by most scholars and physicians for centuries. Leonardo da Vinci drew a sketch of a dissected human fetus in his unpublished notebooks, one of the first such efforts in the history of science. However, in 1651 the great British physician William Harvey (famous for demonstrating the circulation of blood in the body) wrote *On the Generation of Animals*, arguing that all animals come from the egg (*ex ovo omnia* in his original Latin). This argument between spermism and ovism was not finally settled until 1876 when Oscar Hertwig proved that human fertilization occurred when a sperm met an ovum.

The microscope allowed naturalists to see sperm for the first time. One of the founders of microscopy, Antonie van Leeuwenhoek, first described about 30 types of spermatozoa in 1677. Because most scholars thought nature should follow simple mechanical laws, they developed a school of thought called *preformationism*. This idea suggested that there was a tiny version of a human in the head of the sperm (figure 5.1). Known as a *homunculus*, it bore all of the adult features of a human, and it only needed to grow larger through time. Van Leeuwenhoek himself wrote that he could see under the microscope "all manner of great and small vessels, so various and so numerous that I do not doubt that they be nerves, arteries and veins. . . . And when I saw them, I felt convinced that, in no full grown body, are there any vessels which may not be found likewise in semen." Some even postulated that the sperm cells within the testes of the homunculus carried the next generation of homunculi, and so on and so on as they became infinitely tinier and tinier animalcules.

Of course, any decent microscopic view of a sperm shows this is not the case, but the power of suggestion allows people to see things that aren't there— and the quality of the optics of those early microscopes were so poor that they could have seen distorted images and given people the idea that they really could see a tiny human through the lens. Throughout the 1700s, the spermist theory became more entrenched, even as their arguments became more and more convoluted (and virtually untethered by any microscopic data).

The other school of thought, *epigenesis*, also traced its roots all the way back to the ancient Greeks, and it was held by many of the early natural historians, including da Vinci, Gabriele Fallopio (who first described the fallopian tubes), and many more. These people argued that embryos started as very simple cells and gradually grew more complex through time in a series of stages.

The arguments between the preformationists and epigeneticists became more and more heated in the 1700s and 1800s, but it wasn't until better microscopes allowing detailed examination of the structures within cells

Figure 5.1 ▲
A tiny person inside the head of a sperm cell (a homunculus), as drawn by Nicholaas Hartsoeker in 1695. (Courtesy of Wikimedia Commons)

were developed in the early 1880s that modern cell theory was accepted. Along with cell theory came the understanding of what cells are made of and how they grow and change. In the 1760s and 1770s, the German physician Caspar Friedrich Wolff argued that scholars needed to focus on objective descriptions from nature, and not confuse the obscure with theoretical or philosophical considerations. In his view, the microscope observations clearly did not support the idea of a homunculus in every sperm. The final blow came from John Dalton's development of the atomic theory of matter.

It became clear that there could not be a homunculus within the sperm of a larger homunculus, getting smaller and smaller to form infinitely stacked animalcules, because there were lower limits on the size of biological tissues and structures, which were made of even tinier atoms. Finally, in the 1880s biologist Hans Driesch was able to observe how an embryo developed from a sperm and egg in a sea urchin, which had large reproductive cells and embryos and was easier to observe in the lab. These experiments settled the question in favor of epigenesis for good.

The most important figure in the early history of embryology, however, was Karl Ernst Ritter von Baer, Edler von Huthorn. Born of an aristocratic German family in what is now Estonia, his full title included not only the aristocratic "von" and "Edler" but also *Ritter* (knight). Trained in the substandard schools and universities of Estonia, he didn't realize the inadequacies of his medical education until he went to Riga to help the sick and wounded during Napoleon's siege of the city in 1812. Once he graduated from the University of Dorpat in Tartu, Estonia, he went to Berlin, Vienna, and Würzburg to study with the leading scholars of his time. He was introduced to the new field of embryology by Ignaz Döllinger. By 1817, von Baer was a professor at Königsberg University in East Prussia (now Kaliningrad in Russia), a distinguished university going back to 1544 that claims association with many important German scholars and scientists, including Immanuel Kant and Hermann von Helmholtz. There von Baer conducted most of his pioneering embryological research.

In 1834, at the height of his career and fame, von Baer suffered a nervous breakdown and collapse of his health. He decided to change fields, and he gave up embryology and took a new job in St. Petersburg. In his second career, he spent much of his time teaching and doing field research in zoology and geography, including exploring the Caspian Sea and the Arctic Russian island of Novaya Zemlya. He is considered the founder of Russian anthropology and ethnology. In the last few years of his life, he returned to the University of Dorpat and found that his research had inspired many other scientists, including Darwin.

Before he changed careers, von Baer made enormous contributions to embryology. Looking at embryos under his excellent microscope, he discovered and named the blastula stage of development, when the cluster of cells grows into a hollow ball of cells. In 1826 he discovered the mammalian ovum, following in 1827 by the description of the human ovum.

In addition, von Baer followed Wolff's discoveries about epigenetics and developments, and (with Hans Christian Pander) showed that the embryo had three "germ layers": the endoderm, mesoderm, and ectoderm. This culminated with his landmark 1828 book *Entwickelungsgeschichte der Thiere* (Developmental History of Animals), which not only expanded on his discoveries about cells and development but documented years of painstaking research on the development of embryonic birds, mammals, and other animals. In that book, he also described how much embryonic stages of animals—whether they be mammals, birds, reptiles, or amphibians—look like fish.

Von Baer had described something that was already appearing in the work of other embryologists of his time. If you trace adult organisms back to their embryonic stage, they begin to look more and more like each other—and embryos of mammals and birds often look quite fish-like if you go back far enough. In other words, animals repeat or *recapitulate* earlier stages of their past during their development. This idea was popular in the 1790s through the 1820s with many *Naturphilosophes*, including Johann Friedrich Meckel (who described the embryonic cartilage that is the precursor of the bony jaw in vertebrates), Étienne Serres, and Carl Friedrich Kielmeyer. It was formalized by Serres in 1824–1826 and known as the Meckel-Serres law. However, it was not considered evidence of the evolutionary past of the embryonic animals, but just a God-given "pattern of unification" that reflected the unity of nature (just as homology was originally seen before 1859; see chapter 4).

Von Baer was specific in his 1828 book that his embryological observations did not support the Meckel-Serres idea of recapitulation. For example, many features of embryos (especially embryonic organs such as the placenta or the yolk sac) are not features of the adults, so the embryo is not the same as the functioning adult organism. In particular, the embryonic mode of life is very different than the adult mode of life, so the fish-like embryo of a mammal is not an actual fish that could survive as an adult. In addition, there is never complete correspondence between an embryo and the adult of a fish. The chick embryo at one stage may have a heart and circulation like that of a fish, but it lacks most of the other things found in adult fishes. Many of the features that are fixed in adults are only transitory in embryos. Sometimes, parts that should occur later in development appear unusually early, such as the backbone in the chick embryo.

Instead, he posited what are now known as von Baer's laws of embryology. In short, they are:

1. **General characteristics** of the group to which an embryo belongs develop *before* **special characteristics**.
2. **General structural relations** are likewise formed *before* the most **specific** appear.
3. The form of any given embryo does not converge upon other definite forms but separates itself from them.
4. The **embryo** of a higher animal form never resembles the **adult** of another animal form, such as one less evolved, but only its embryo.

The basic thrust of von Baer's laws is that all vertebrate embryos start out as very generalized unspecialized vertebrates, later adding more specializations that distinguish them as adults. In his words, "the further we go back in the development of vertebrates, [the] more similar we find embryos both in general and in their individual parts. . . . Therefore, the special features build themselves up from a general type."

The ideas that von Baer and others had promoted since the 1820s and 1830s were widely discussed by many naturalists, including one of Darwin's favorite professors, Robert Grant, during his early medical education at the University of Edinburgh. As Darwin was formulating his ideas about the "species problem" in 1842, he ran across Johannes Müller's summary of the embryological research by von Baer. From reading and the ideas Darwin learned from Grant, Darwin was already well aware that embryos of organisms (such as the barnacles he studied) often showed features that helped diagnose and distinguish otherwise identical adults. They also showed the origins of some of the organs that were known to be homologous (see chapter 4). Now he realized that the old ideas of recapitulation from Meckel and Serres perfectly fit his concept of a common ancestry of all animals. In Darwin's unpublished notebooks, where he wrote his thoughts about evolution, we can see how the ideas of the embryologists were influencing his thinking. In the "B Transmutation Notebook," he quoted Geoffroy as saying, "generation is a short process by which one animal passes from worm to man highest or typical of changes which can be traced in the same organ in *different* animals in scale."

The embryological suggestion of descent from a common ancestry was one of the most powerful and persuasive pieces of evidence Darwin could muster in 1859. He wrote an extended section about the topic in *On the Origin of the Species* (see the epigraph at the beginning of this chapter). As he wrote to his friend, American botanist Asa Gray, in 1860, "embryology is to me by far the strongest single class of facts in favor of change of forms." In his autobiography, Darwin wrote, "hardly any point gave me so much satisfaction when I was at work on the Origin as the explanation of the wide difference in many classes between the embryo and the adult animal, and of the close resemblance of the embryos within the same class." In the fifth edition of *On the Origin of Species*, Darwin added "historical sketches" that credited the naturalists who had influenced him. There he wrote, "Von Baer, towards whom all zoologists feel so profound a respect, expressed about the year 1859 . . . his conviction, chiefly grounded on the laws of geographical distribution, that forms now perfectly distinct have descended from a single parent-form."

For his part, von Baer (who was by then retired but still at the University of Dorpat, where he survived until 1868 and died at age 84) was not too happy with Darwin's use of his discoveries. He did not object to the idea of transmutation of species, and he agreed that embryos showed how all vertebrates had a common ancestry. In fact, some of his writings show one of the first "family trees" of life based on embryos. But von Baer was from an older, more mystical school of *Naturphilosophie* and didn't like the impersonal mechanistic implications of natural selection. Instead, he believed animals were striving and changing to reach a higher goal (known as teleology), and that there were mysterious internal forces in nature that directed this kind of evolution.

Darwin and von Baer were rather cautious in their use of embryological transformations as support for the idea of evolution. Not so for the young German biologist Ernst Haeckel. Born in Potsdam in 1834, and son of a government lawyer, he got his medical degree in 1858 but discovered he didn't enjoy being around suffering patients. So he focused his attention on biological and medical research instead. He worked with the legendary embryologist Karl Gegenbaur at the University of Jena for three years and earned a teaching degree in anatomy and zoology in 1861. He ended up remaining at Jena for the rest of his 47-year career, where he made his reputation working on sponges, worms, and other invertebrates. He was especially known for studying and naming hundreds of species of the tiny siliceous amoebas known as radiolarians, which were hugely abundant in the ocean plankton. Some of the radiolarian and sea jellies and other marine creatures had been sent to him for study from the first great oceanographic voyage of the HMS *Challenger* in 1874–1876 because he was one of the only scholars in the world who knew radiolarians or sea jellies.

Haeckel was also a "big ideas" person, often making sweeping generalizations and predictions. For example, Haeckel predicted that the origin of humans would be found in Southeast Asia, inspiring Eugene Dubois to go to the Dutch East Indies and find "Java man." Haeckel even gave a name to this as yet undiscovered fossil, *Pithecanthropus alalus* (ape man without speech), and Dubois named his specimens *Pithecanthropus erectus* after they were described. (Today they are called *Homo erectus*.) Haeckel was also fond of creating new words, such as *ecology, phylum, phylogeny, ontogeny*, and *Protista*, to reflect his grand synthetic ideas about biology. In addition, Haeckel was the first to publish an explicit tree of life, illustrating Darwin's idea that all life has a common ancestry.

As a big ideas scientist, Haeckel was immediately captivated by Darwin's book when he finally read it in 1864. During his 1866 expedition to the Canary Islands, he took a side trip to England and visited Charles Darwin at his estate Down House, as well as Thomas Henry Huxley and Charles Lyell. When Haeckel settled down and became established at Jena, he soon became Darwin's chief advocate in Germany, pushing hard to update science education and get evolution established in all the textbooks. Haeckel's 1868 work *Natürliche Schöpfungsgeschichte* (Natural Creation History) was a best-selling popular explanation of evolution, especially after it was translated into English in 1876 as *The History of Creation*.

It was in the field of embryology, however, that Haeckel made his biggest contribution to evolution. He wrote a massive textbook, *Generelle Morphologie*, in 1866 that synthesized Darwin's ideas with the older recapitulationist ideas of Meckel and Serres. Ignoring the cautious approach of von Baer, he went full tilt into the issue and insisted that embryonic development completely repeats the evolutionary history of organisms. In the words he coined himself, "ontogeny recapitulates phylogeny" or "embryonic history repeats evolutionary history." He even argued that the as yet undiscovered common ancestors of many of the groups on his "tree of life" would look exactly like the most primitive embryos of various living organisms. This argument did not hold up well, and most people regard it as one of Haeckel's major mistakes. But he was on the right track with his tree of life, even if he overextended the embryological evidence. More important, he ignored von Baer's cautions that many embryonic features are unique to embryos and have nothing to do with the adult forms of any animal.

Finally, he published diagrams emphasizing the obvious similarities between the fish-like earliest embryonic stages of most vertebrates and showing how they all developed into different kinds of adults. In this area, he may have been a bit overzealous in his drawings (figure 5.2). Some of

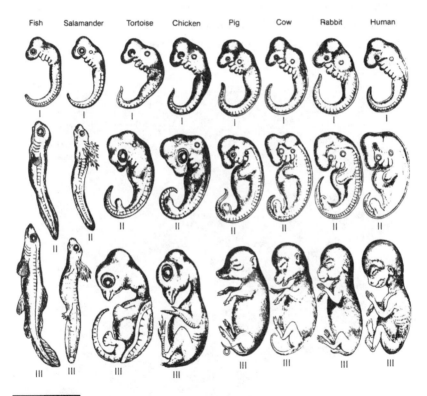

Figure 5.2 ▲

Diagram of embryonic development of the vertebrates, showing the similarity of fish-like features of early embryos, becoming more specialized as it differentiates into a fish, a reptile, or a mammal. Modified from Haeckel's problematic illustration, this diagram is fundamentally correct, although Haeckel might not have been totally accurate in every detail. (From George Romanes, *Darwin and After Darwin* [Open Court, Chicago, 1910])

them were not entirely accurate and made the embryos look more fish-like than they actually did. In one case, he used the same illustration to show a dog, a chick, and a turtle embryo, making them appear identical; he later had to fix this mistake and show the real embryos. They do look extremely similar, but Haeckel's inaccurate drawings made the entire argument look bad in retrospect. Anti-evolutionists have raised this criticism over and over again, arguing that embryology does not support evolution. However, if you look at any good series of images showing embryonic development of vertebrates, the evidence is obvious, no matter what mistakes Haeckel may have made over 150 years ago.

Many people never think much about what embryology tells us about our evolutionary history. But if we are not descended from a common ancestor with fishes, reptiles, and other mammals, why do we have their distinctive features during our development? Figure 5.3 shows a human embryo five

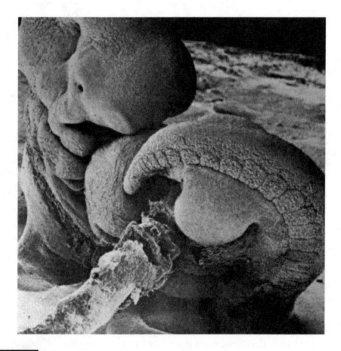

Figure 5.3 ▲

This is what you looked like five weeks after conception. You still had many fish-like features, such as a well-developed tail and the embryological precursors of gill slits, both of which are lost in most human embryos as they develop. (From the IMSI Photo Library)

weeks after fertilization. Our well-developed tail is still prominent, as is the embryological precursor to the gill slits. As von Baer pointed out, this embryo is not the same as an adult fish and could not live as a swimming adult, but it is similar to an embryonic fish at the same stage of development. If we did not have common ancestors with fish (and with reptiles and other mammals, all of whose embryos look extremely similar to ours in the earliest stages), why do we have such strong embryonic similarity? This argument is as powerful today as it was when Darwin used it in 1859.

FOR FURTHER READING

Gilbert, Scott F., and Michael J. F. Barresi. *Developmental Biology*. 11th ed. Sunderland, Mass.: Sinauer, 2016.

Gould, Stephen Jay. *Ontogeny and Phylogeny*. Cambridge, Mass.: Harvard University Press, 1977.

Hall, Brian K. *Evolutionary Developmental Biology*. Dordrecht, Germany: Springer, 1999.

THE SINKING OF NOAH'S ARK

In considering the distribution of organic beings over the face of the globe, the first great fact which strikes us is that neither the similarity nor the dissimilarity of the inhabitants of various regions can be wholly accounted for by climatal or other physical conditions.

—CHARLES DARWIN, *ON THE ORIGIN OF SPECIES* (1859)

Before the 1700s, naturalists had very little knowledge of the animals that lived outside Europe. The early classifications by people such as Linnaeus (see chapter 7) listed only 4,200 species of animals, nearly all from Europe or the Middle East, but occasionally including exotic animals imported from Asia, Africa, or the New World. According to Genesis 6, all animals migrated from Mt. Ararat (now in Turkey) to their present locations. This explanation was first found in the ancient myths of the Middle East, at a time when people knew of barely 100 species of animals in their limited region. In addition to their domesticated cattle, sheep, goats, pigs, dogs, cats, and horses, they knew of very few wild animals, so it was plausible to believe that a few hundred of them could be brought onto Noah's ark.

The myth of a worldwide flood destroying all of humanity goes back to *The Epic of Gilgamesh*, which dates to about 2750 BCE. The Sumerian hero Ziusudra (called Atrahasis by the Akkadians and Utnapishtim by the Babylonians) was warned by the earth goddess Ea to build a boat because the god Ellil was tired of the noise and trouble of humanity and planned to wipe them out with a flood. When the floodwaters receded, the boat was grounded on the mountain of Nisir. After Utnapishtim's boat was stuck for seven days, he released a dove, which found no resting place and returned. He then released a swallow that also returned, but the raven that was released the next day did not return. Utnapishtim then sacrificed to Ea on the top of Mount Nisir. The story is nearly identical to that of Noah's flood, not only in its plot and structure but also in the details of its phrasing. Only the characters' and the gods' names and a few details were changed to suit the differences between the monotheistic Hebrew culture and the polytheistic cultures of the Sumerians, Akkadians, and Babylonians.

Two centuries of biblical scholarship have shown that the Hebrew Torah (i.e., the first five books of the Old Testament) was a composite of different sources written by different groups of people at different times. Thus there are many contradictions, such as Genesis 7:2 (from one group of priests known as the J source) saying that Noah took seven pairs of each clean beast in the ark, but Genesis 7:8–15 (from a different group of priests known as the P source) said he took only one pair of each beast in the ark. In Genesis 7:7, Noah and his family finally enter the ark, and in Genesis 7:13 they enter it all over again (the first verse from the J source, the second from the P source). Of course, the ancient Hebrews did not take the Torah literally. They used it as a guide to understanding their relationship to Yahweh and saw no problems with the story of Noah's ark. Their world was limited to the Tigris-Euphrates valley and parts of the eastern Mediterranean, and only a few hundred species of animals lived in that region. Today millions of species are known, and the idea that they would all fit on one boat is comically absurd.

Even as Linnaeus recognized as many as 4,200 species of animals (already a problem for Noah's ark), an array of voyages and scientific expeditions in the years after 1700 led to many discoveries that demolished this view of biogeography. Far-off places such as South America, Africa, Madagascar, Southeast Asia, Australia, and the Pacific islands yielded huge numbers of strange species of animals and plants, completely upending the Eurocentric view of nature. Tropical regions, in particular, had much richer and more unusual fauna and flora than the depleted wildlife of northern Europe, already severely impacted by its large human populations, widespread agriculture, and long history of deforestation. Many formerly common European species, including wild cattle (aurochs) and wild horses, had been driven to extinction by the early Holocene (ca. 10,000 years ago). In contrast, the tropics were so rich and diverse that natural historians were overwhelmed trying to describe and catalog all these new species from exotic lands.

More important, it became apparent that these geographic distributions could not be explained by the Noah's ark story. Why did the continent of Australia have fauna dominated by pouched mammals (marsupials) and no native placental mammals? Did the marsupials run straight from Mt. Ararat toward Australia, but placentals didn't even try to get there? Even more striking was the fact that many of the marsupials had body forms that mimicked the placentals occupying a similar ecological niche on other continents (figure 6.1). There were marsupial equivalents of wolves, badgers, cats, flying squirrels, groundhogs, anteaters, moles, rabbits, and mice—yet all were pouched mammals unrelated to their placental counterparts on other continents. We now view this as an outstanding example of convergent evolution, but these major discoveries were made a century before evolution explained them.

There is a famous saying that "travel is broadening," and it was certainly a factor in the discovery of biogeographic distributions of animals and plants. It changed the perspective of both Darwin and the codiscoverer of natural selection, Alfred Russel Wallace. Darwin was relatively well educated in British natural history for a young man just out of Cambridge when he left on the *Beagle* voyage in 1831, but the five years sailing around the world, visiting many exotic locations such as the Brazilian rain forests, the Galápagos Islands, plus Australia and Africa, completely transformed his conception of the world of animals and plants. He was only 22 when the HMS *Beagle* first reached South America. As a typical Englishman, he was used to cool, dreary, and wet weather most of the time and just a handful of wild animals in the largely domesticated landscape of England. Darwin's first stop in South America was on February 28, 1832, and he disembarked in Salvador, Bahia, Brazil, at the edge of the Amazon rain forest. He was immediately captivated and overwhelmed by the richness of the wildlife and the intensity of the tropical jungle foliage, the stifling heat and humidity, and the numerous biting insects he encountered. As he wrote later, he was rapturous over "the elegance of the grasses, the novelty of the parasitic plants, the beauty of the flowers." He walked around in a semidazed state, almost unable to take in all the new sights, sounds, smells, and thoughts: "To a person fond of natural history, such a day as this brings with it deeper pleasure than he can ever hope to experience again." Each stop in the Brazilian rain

Placentals

Marsupials

Wolf
(Canis)

Tasmanian wolf
(Thylacinus)

Ocelot
(Felis)

Native cat
(Dasyurus)

Flying squirrel
(Glaucomys)

Flying phalanger
(Petaurus)

Ground hog
(Marmota)

Wombat
(Phascolomys)

Anteater
(Myrmecophaga)

Anteater
(Myrmecobius)

Mole
(Talpa)

Mole
(Notoryctes)

Mouse
(Mus)

Mouse
(Dasycercus)

Figure 6.1 ▲

The native fauna of Australia consists mainly of pouched marsupials, which have con-verged remarkably on their placental counterparts from other continents, even though the two groups are not closely related. In Australia, there are marsupials that look vaguely like wolves, cats, flying squirrels, groundhogs, anteaters, moles, and mice—but they are all pouched mammals. (Modified from George Gaylord Simpson and William Beck, *Life: An Introduction to Biology*, 2nd ed. [New York: Harcourt, Brace, & World, 1965])

forest was ecstasy to a hard-core naturalist and collector like Darwin. One day he collected no less than 69 different species of beetle, all new to science. Darwin wrote, "it is enough to disturb the composure of the entomologist's mind to contemplate the future dimension of a complete catalogue."

On April 3, the *Beagle* reached Rio de Janeiro, where he rode with a party of Englishmen going to a coffee plantation about 100 miles inland. It was a difficult journey, with the blazing heat, poor accommodations, and vampire bats that bit their horses at night, but Darwin was still rapturous in his amazement. Brilliant birds and butterflies were everywhere, and hummingbirds darted from flower to flower. Cabbage palms towered 50 feet overhead, and long lianas hung down from them. Some forests looked positively prehistoric, with enormous tree ferns that were relicts of the time before the dinosaurs. The giant tree canopy towering overhead reminded him of a huge high-ceilinged cathedral, with just small shafts of light penetrating through the thick covering of leaves.

Darwin continued to be staggered by the beauty and abundance of the Brazilian jungle. As he wrote in *The Voyage of the Beagle* (1836),

> It was impossible to wish for any thing more delightful than thus to spend some weeks in so magnificent a country. In England any person fond of natural history enjoys in his walks a great advantage, by always having something to attract his attention; but in these fertile climates, teeming with life, the attractions are so numerous, that he is scarcely able to walk at all.

As discussed in chapter 3, the huge biological diversity in Brazil was a contrast with the limited number of weird animals on the Galápagos Islands. By the time Darwin had sorted out the mockingbirds, and the subsequent discovery that the "grosbeaks," "wrens," and other birds were actually modified finches, Darwin realized that the only explanation for their diversity was not individual creation on each island but diversification from an ancestral finch population that had blown in from South America. In fact, the geographic evidence was so important to Darwin that he devoted two whole chapters to it in *On the Origin of Species* in 1859, including this:

> The most striking and important fact for us in regard to the inhabitants of islands, is their affinity to those of the nearest mainland, without being actually the same species. [In] the Galapagos Archipelago . . . almost every product

of the land and water bears the unmistakable stamp of the American continent. There are twenty-six land birds, and twenty-five of these are ranked by
Mr. Gould as distinct species, supposed to have been created here; yet the
close affinity of most of these birds to American species in every character,
in their habits, gestures, and tones of voice, was manifest. . . . The naturalist,
looking at the inhabitants of these volcanic islands in the Pacific, distant several hundred miles from the continent, yet feels that he is standing on American land. Why should this be so? Why should the species which are supposed
to have been created in the Galapagos Archipelago, and nowhere else, bear
so plain a stamp of affinity to those created in America? There is nothing in
the conditions of life, in the geological nature of the islands, in their height
or climate, or in the proportions in which the several classes are associated
together, which resembles closely the conditions of the South American coast:
In fact there is a considerable dissimilarity in all these respects. On the other
hand, there is a considerable degree of resemblance in the volcanic nature of
the soil, in climate, height, and size of the islands, between the Galapagos and
Cape de Verde Archipelagos: But what an entire and absolute difference in
their inhabitants! The inhabitants of the Cape de Verde Islands are related to
those of Africa, like those of the Galapagos to America. I believe this grand
fact can receive no sort of explanation on the ordinary view of independent
creation; whereas on the view here maintained, it is obvious that the Galapagos Islands would be likely to receive colonists, whether by occasional means
of transport or by formerly continuous land, from America; and the Cape de
Verde Islands from Africa; and that such colonists would be liable to modification—the principle of inheritance still betraying their original birthplace.

This quote also raises a second point that Darwin noted: The fauna and
flora of oceanic islands are often very odd and unbalanced. Most of the
larger islands have their own unique species of animals and plants that are
very different from those on the mainland; these are known as *endemic* species. These include not only Darwin's Galápagos finches and mockingbirds
and tortoises but also the odd life found on nearly every island. Madagascar
is famous for its weird collection of lemurs, predatory cat-like fossas, insec-
tivorous tenrecs, and a whole suite of unique birds (60 percent endemic to
Madagascar), reptiles (90 percent endemic), amphibians, and two whole
endemic families of fish, as well as lots of endemic insects and 100 percent of its 651 species of land snails. And there are almost 15,000 species of

plants, more than 80 percent of which are found nowhere else. In total, 90 percent of its species are unique to Madagascar.

New Zealand is unique in that it never had any land mammals (except bats), so birds, reptiles, and other animals perform the role of mammals elsewhere. In addition, there are few predators, so many of its animals had not adapted antipredatory behaviors (until humans and their animals arrived about 900 years ago). It is home not only to the kiwi but also to the huge nocturnal flightless parrot known as the kakapo and the predatory parrot called the kea, gigantic flightless moas, and many other unique species not found elsewhere. Almost every island around the globe has a similar list of unique, endemic species found only there, yet many species had distant relatives among the animals from the nearest mainland. If they had all been created by God at the same time and moved to these islands after leaving Mt. Ararat, why were the inhabitants of each island unique? If their ancestors escaped from the mainland, had arrived there long ago, and were isolated from competition from mainland animals, it only makes sense that they evolved to inhabit these new niches.

The further the island was from a continent, the more extreme the peculiarity of the fauna. Darwin did not study the creatures of Hawaii, but these islands are a long way from any land mass. The only way creatures could reach these islands is by flying or being blown there over huge stretches of the Pacific Ocean. Consequently, Hawaii has no native land mammals (only bats), no reptiles or amphibians, and only a few native freshwater fish (all of whom can also swim in salt water). It is ruled by the descendants of those creatures blown there by accident during typhoons: an entirely unique assemblage of 71 species of birds found nowhere else, some of which show huge evolutionary diversification (such as the Hawaiian honeycreepers, birds with long bills for sipping nectar, as well as unique endemic ducks, finches, coots, rails, hawks, and others), lots of endemic insects (including several unique species of fruit fly found nowhere else), and hundreds of species of endemic plants. It makes no sense that these creatures had somehow walked from Mt. Ararat, but it is clearly consistent with the idea that only a few lucky survivors were blown to these most remote islands in the world and diversified into an array of endemic species in the absence of any competition from mainland species.

Wallace, the codiscoverer of natural selection, had a very different career path than that of Charles Darwin (figure 6.2). Born in 1823 in

Figure 6.2 ▲

Photograph of Alfred Russel Wallace as a young man. (Courtesy of Wikimedia Commons)

Llanbadoc, Wales, of an impoverished family of nine children, he was 14 years younger than Darwin and looked on Darwin as his older mentor. Wallace's poverty meant that his formal schooling was brief, and he scratched out a living for many years as a surveyor, mapmaker, and a schoolteacher. Through his rambles surveying and mapping the countryside, he became an enthusiastic naturalist, especially in collecting insects. Inspired by Darwin's book about the *Beagle* voyage and accounts of Alexander von Humboldt and other explorers, in 1848 the 25-year-old Wallace and his friend William Henry Bates (soon to become a famous naturalist himself) set off on a hazardous collecting trip to the Amazon jungle. They spent four years there and collected large numbers of new species of animals (especially insects) to sell to the voracious market of exotic natural history collectors in England. Tragically, on the way home, Wallace's ship caught fire and sank, destroying almost all of Wallace's valuable collections. It left him and the crew to float in an open lifeboat for 10 days before they were rescued. When he returned to London, Wallace spent 18 months living off the insurance payments for his lost collections and the sale of specimens he had shipped earlier. From 1854 to 1862, he collected specimens and explored the Malay Archipelago (now Indonesia and Malaysia), nearly dying of malaria or from accidents several times. By the end of this expedition, he had seen and documented more new animals and plants than any human alive at that time, and he had a wealth of experience in two of the world's most diverse tropical regions, which gave him total command of the patterns of biogeographic distribution of animals.

While staying in Ternate in the Malaku Islands in 1858, Wallace endured a severe bout of malaria and was near death many times. During lucid moments, he wrote down his own version of natural selection and mailed it to (of all people) Charles Darwin. Famously, Darwin was shocked when he received it and was worried that he had been scooped after 20 years of anxiety and procrastination in publishing his original ideas on natural selection, which he had jotted down in 1838. Darwin appealed to his friends, geologist Charles Lyell and botanist Joseph Hooker, for an honorable solution to his dilemma. They arranged to have both Wallace's letter and two of Darwin's early sketches about natural selection read at a meeting of the Linnaean Society, so they would share credit for the idea. No one apparently thought much of it at the time. Thomas Bell, president of the society at the session in 1858 when the Darwin-Wallace documents were read, commented that "the year which has passed has not, indeed, been marked by any of those striking discoveries which at once revolutionize, so to speak, the department of science on which they bear." He could not have been more wrong! Clearly, no one understood the importance of the Darwin-Wallace papers at the time.

Meanwhile, Darwin realized that the idea was in the air and that he would be scooped if he didn't act fast to establish how much more he had done on the problem than anyone else. He feverishly wrote a "short" version of his long-delayed book on the topic, and it sold out on the day it was published in November 1859.

Returning to London in 1862, Wallace learned of all the excitement around Darwin's book in the previous three years, but he never begrudged Darwin for becoming famous for an idea that they both had come up with independently. Wallace became one of Darwin's staunchest defenders without drawing any attention to his own discovery of the idea. Indeed, one of his last books in 1889 on the topic was called *Darwinism*.

Wallace contributed to the debate over evolution, and his enormous global experience with exotic animals made him a pioneer in the subject of biogeography. He virtually established the entire field with his publication of *The Geographical Distribution of Animals* in 1876. In that work, he brought together all the new information on faunal provinces and formalized many of the ideas of biogeography that persist today. One of his most famous contributions was the discovery of the boundary in the Indonesian archipelago between islands that have mostly Asian-influenced faunas (such as tapirs, rhinos, tigers, and other Asian jungle animals) and those that are dominated by Australian animals (such as wallabies and other marsupials, egg-laying "spiny anteaters" or echidnas, and large flightless birds like the cassowary). This boundary came to be known as "Wallace's line" (figure 6.3), and today we know it as a striking example of how two distinct faunas can mingle once one continent (Australia) drifts into the influence of another continent (Asia).

Most of biogeography prior to the 1960s focused on how the animals and plants of one geographic region were unique to that region and found nowhere else. This pattern made the Noah's ark story implausible because animals were not distributed in a pattern radiating out from Mt. Ararat in Turkey. Even more revealing was the way in which animals from different regions were similar and even closely related. A striking example of this is the primitive group of flightless birds known as ratites, which includes the ostrich in Africa, the rhea in South America, the cassowary and emu in Australia and New Guinea, and the kiwi in New Zealand. Their distribution on all the southern continents was long a mystery, especially because they could not fly across the oceans that separate those continents. In the 1960s, when plate tectonics established that all of the southern continents were once part of a supercontinent called Gondwana that broke up late in the Cretaceous Period, the pattern finally made sense (figure 6.4). These birds had originated in Gondwana before it broke up and remained in their homelands, evolving and diverging into different kinds of ratites, such as ostriches and rheas and emus. (The story is somewhat complicated by the fact that one fossil ratite is known from Europe about 40 million years ago.)

Likewise, the pouched mammals, or marsupials (see figure 6.1), have distribution that is a remnant of their original Gondwana range. These include the familiar kangaroos, wallabies, bandicoots, wombats, koalas, and Tasmanian devils of Australia as well as a number of primitive marsupials in South America and one fossil marsupial in Antarctica. In addition,

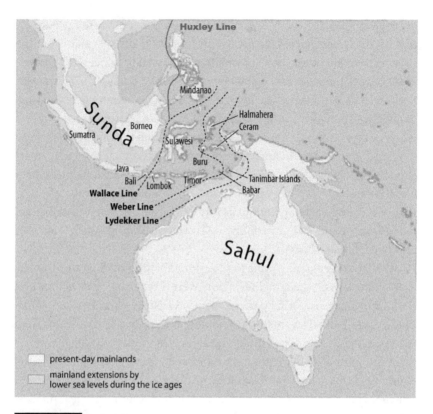

Figure 6.3 ▲

Alfred Russel Wallace collected in many places in the Malay Archipelago of the Dutch East Indies (now Indonesia and Malaysia) and got to know the differences in the faunas of each island well. He was struck by the similarity of the animals of New Guinea and the western islands with Australia, especially typical Australian marsupials (such as wallabies), egg-laying echidnas, or spiny anteaters, and large flightless birds like the cassowary. At the western end of the island chain, most of the mammals resembled those of mainland Southeast Asia (such as tapirs, rhinos, tigers, and other Asian jungle animals). He drew a line between Borneo and Sulawesi, extending down to Bali and Lombok, where the transition seemed to be most striking. This line marks the deepest part of the straits (such as the Lombok Strait) that separate the islands (and continental shelves) of Asia from those of Australia. Even when sea level dropped 400 feet during the last Ice Age, there was always a deepwater separation between these islands, making it difficult for land animals (tigers and rhinos to the east, and marsupials and monotremes to the west) to cross the barrier. Since then, geologists and biologists have proposed other lines of demarcation between faunal provinces. In modern terms, these lines represent the transition zone that has gradually diminished since the Cretaceous as the Australian plate drifts toward the Asian plate, making their faunas more and more similar. (Courtesy of Wikimedia Commons)

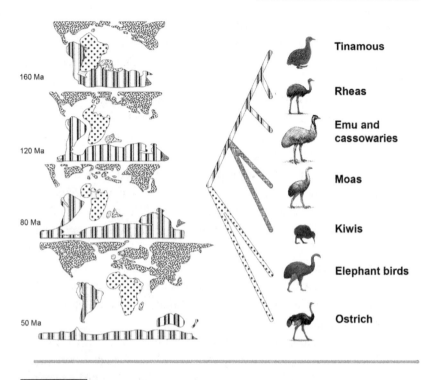

Figure 6.4 ▲

The branching sequence of evolution of the large flightless ratite birds as explained by the time series breakup of the Gondwana into continents, leaving one or two ratites on each continent today. (Ma = millions of years ago) (Redrawn from several sources)

marsupials used to be much more dominant in South America than they are now, with a great radiation of extinct predatory marsupials that resembled wolves, hyenas, and even a saber-toothed marsupial that closely paralleled the saber-toothed cats of the northern continents. Only the relatively primitive opossums managed to escape to the northern part of the world during the Cretaceous, but they never came to dominate the northern continents as marsupials did in Australia and South America, where there was relatively little or no competition from placental mammals that ruled elsewhere.

A similar relict Gondwana pattern is known from the primitive side-necked turtles, or pleurodires, who bend their necks sideways and fold their neck and head under the front edge of the shell for protection (rather than pull their head inside their shell, as the more familiar cryptodire turtles do).

Today they are found only in Africa, Madagascar, Australia, and South America, but fossils have been found in India as well, so they were on every Gondwana continent except Antarctica (which has a poor fossil record due to the ice cap covering most of the rocks). In the past, they occasionally spread to Eurasia and North America, but most of their evolution took place in Gondwana before it broke up. Other examples can be found in two families of frogs, the Microhylidae and Natatanura, found only on the Gondwana remnants today. Because amphibians cannot cross saltwater barriers due to their porous skins, the only explanation for their distribution around the southern continents is divergence before Gondwana broke up in the Late Cretaceous. The three living species of lungfish are found in Africa, South America, and Australia, although they are remnants of a once worldwide distribution of lungfish during the Age of Dinosaurs. Similar examples can be found in many other groups of animals (especially insects and spiders) and plants. A striking example is the southern beech (*Nothofagus*), found only on the cooler and wetter parts of Gondwana (New Zealand, Tasmania, and Patagonia) today.

We have come a long way from the ancient myth of Noah's ark to the modern biogeography pioneered by Darwin, Wallace, and many others. Not every pattern of distribution of animals and plants is fully understood even today. But most of it only makes sense in light of evolution and the other great scientific revolution of this century—plate tectonics.

FOR FURTHER READING

Brown, James H., and Arthur C. Gibson. *Biogeography*. St. Louis, Mo.: Mosby, 1983.

Browne, Janet. *The Secular Ark: Studies in the History of Biogeography*. New Haven, Conn.: Yale University Press, 1983.

Cox, C. Barry, and Peter D. Moore. *Biogeography: An Ecological and Evolutionary Approach*. 7th ed. Cambridge, Mass.: Blackwell, 2005.

Darlington, Peter J. *Zoogeography: The Geographical Distribution of Animals*. New York: Wiley, 1957.

Lomolino, Mark V., Brett R. Riddle, Robert J. Whittaker, and James H. Brown. *Biogeography*. 4th ed. Sunderland, Mass.: Sinauer, 2010.

McCarthy, Dennis. *Here Be Dragons: How the Study of Animal and Plant Distributions Revolutionized Our Views of Life and Earth*. New York: Oxford University Press, 2009.

Morrone, Juan J. *Evolutionary Biogeography: An Integrative Approach with Case Studies.* New York: Columbia University Press, 2008.

Parenti, Lynne R., and Malte C. Ebach. *Comparative Biogeography: Discovering and Classifying Biogeographical Patterns of a Dynamic Earth.* Berkeley: University of California Press, 2009.

Pielou, E. C. *Biogeography.* New York: Wiley-Interscience, 1979.

Simpson, George Gaylord. *Evolution and Geography: An Essay on Historical Biogeography with Special Reference to Mammals.* Eugene: Oregon State System of Higher Education, 1962.

THE BRANCHING TREE OF LIFE

God created, but Linnaeus classified.

—CAROLUS LINNAEUS, 1758

The ancient Greeks, such as Aristotle, knew of about 550 different kinds of animals in their day. They tried to make sense of all these animals, creating schemes to group similar things together and classify them. Some grouped organisms on properties that humans favored (good to eat, eat only in emergency, inedible, or poisonous) or on properties of their ecology (for example, most animals in the ocean were called "fish," including "starfish" and "shellfish" and whales). By the early 1700s, the more than 6,000 recognized species of plants and 4,200 species of animals had been organized into a great confused and conflicting mess of classification schemes proposed by natural historians. Most of these classifications were arbitrary and highly unnatural (for example, flying fish and birds were put together because they both fly, or turtles and armadillos because of their armor), and everybody had their own favorite scheme.

The classification method that eventually prevailed was proposed by the Swedish botanist Carl von Linné, known to us by his Latinized name, Carolus Linnaeus (all scholars of his day wrote in Latin). As a botanist, Linnaeus recognized that the most fundamental and diagnostic properties of plants are found in their reproductive structures, particularly their flowers. His "sexual system" for classifying plants was published as *Species Plantarum* in 1752, and it created a scandal because of its sexual overtones. Eventually it won out over all of the competing systems because flowers are clearly more useful in determining the true relationships of flowering plants than are any leaves or stems. Linnaeus tried a similar approach for animals, using fundamental structures (such as hair and mammary glands in mammals) rather than superficial ones (such as flight or armor). His *Systema naturae, regnum animale* (The System of Nature, Animal Kingdom) was first published in 1735, and its tenth edition (1758) is now regarded as the starting point of modern classification.

Linnaeus's original classifications became outdated as thousands of new species were described after 1758, but his fundamental system still survives. Each species is given a binomen (two-part name), consisting of the genus (plural, genera) name (always italicized or underlined, and always capitalized) and the trivial name indicating the species (always italicized or underlined but never capitalized). For example, our genus is *Homo* ("human" in Latin) and our trivial name is *sapiens* ("thinking" in Latin), so our species name is *Homo sapiens* (abbreviated *H. sapiens*). Genera are then grouped into higher categories: family, order, class, phylum (plural, "phyla"), and kingdom. For example, humans are members of the kingdom Animalia (there are also kingdoms for plants, fungi, and single-celled organisms), the phylum Chordata (including all other backboned animals), the class Mammalia (mammals), the order Primates (including lemurs, monkeys, apes, and ourselves), the family Hominidae (including our own genus and the extinct *Australopithecus, Sahelanthropus, Orrorin, Ardipithecus,* and *Paranthropus*), the genus *Homo* (including other extinct species such as *Homo habilis* and *H. erectus*), and our species *H. sapiens*. Notice that this classification scheme is hierarchical (figure 7.1). Each rank is grouped into larger ranks, so there may be several species in a genus, several genera in a family, and so on.

Linnaeus and his contemporary natural historians viewed their task as a religious mission. They thought that deciphering the "Natural System" of life would reveal the workings of the mind of the Creator that set up this Natural System. But the obvious clusters of organisms into groups within groups suggested something else to Charles Darwin. This hierarchical, nested, branching structure of life only made sense if life had descended

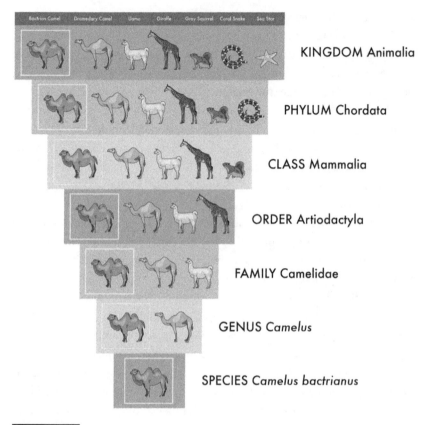

Figure 7.1 ▲

The hierarchy of classification showing how each rank or group is nested within a larger one. In this case, the species of the two-humped Bactrian camel (*Camelus bactrianus*) is one of two species within the genus *Camelus*, which is one of several genera within the family Camelidae, which is a group within the order Artiodactyla, which is one of several orders within the class Mammalia, which is one of several orders within the phylum Chordata, one of many phyla in the kingdom Animalia.

from a common ancestry in a branching fashion (figure 7.2). As Darwin wrote in *On the Origin of Species* in 1859:

> From the most remote period in the history of the world, organic beings have been found to resemble each other in descending degrees, so that they can be classed in groups under groups. This classification is not arbitrary like the groups of the stars in constellations. The existence of groups would have been

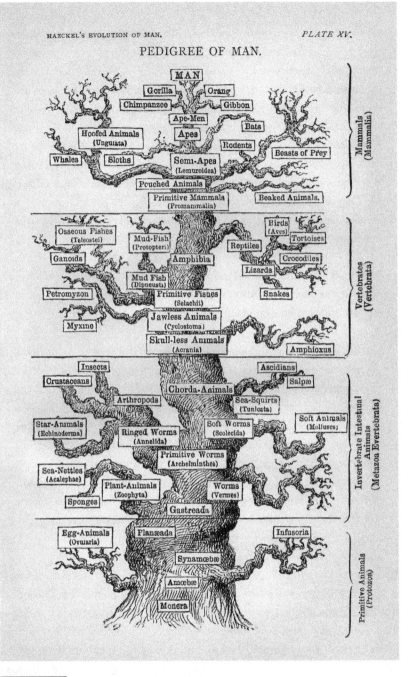

Figure 7.2 ▲

Ernst Haeckel's (1874) evolutionary tree, one of the first graphic representations of phylogeny as a "tree of life" ever to appear in print. Although many of the branches are real animals, the first five stages (monera, amoeba, synamoeba, planaea, and gastraea) are reconstructed based on the ontogeny of higher forms and not on living organisms then known to Haeckel. (From E. Haeckel, *Anthropogenie* [Leipzig, Germany: W. Engleman, 1874])

of simple significance if one group had been exclusively fitted to inhabit the land, and another the water; one to feed on the flesh, another on vegetable matter, and so on; but the case is widely different, for it is notorious how commonly members of even the same subgroup have different habits. Naturalists try to arrange the species, genera, and families in each class on what is called the Natural System. But what is meant by this system? Some authors look at it merely as a scheme for arranging together those living objects which are most alike, and separating those most unlike. Many naturalists think something more is meant by the Natural System; they believe it reveals the plan of the Creator. I believe that a community of descent—the one known cause of close similarity in organic beings—is the bond which is partially revealed by our classifications.

Although Linnaeus had not intended to provide evidence for evolution, a century later his classification scheme became one of Darwin's best arguments. As Stephen Jay Gould wrote in 2000:

Linnaeus's taxonomic scheme designates a rigorously nested hierarchy of groups (starting with species as the smallest unit) embedded within successively larger groups (species within genera within families within orders and so forth). Such a nested hierarchy implies a single branching tree with a common trunk that ramifies into ever finer divisions of boughs, limbs, branches, and twigs. This treelike form just happens to express the hypothesis that interrelationships among organisms record a genealogical hierarchy built by evolutionary branching. Linnaeus's system thus embodies the causality of Darwin's world. Linnaeus's creationist account just happened to imply a structure that, by pure good fortune, could be translated without fuss or fracture into the evolutionary terms of Darwin's new biology. (18)

In doing so, Darwin changed the goals of classification. It was no longer just a nice but arbitrary system of arranging things into pigeonholes. Taxonomy now had an evolutionary meaning as well, and taxonomists were trying to create natural groups that reflected evolutionary history. Although these goals are not contradictory, they do not always agree either. Some taxonomists view organisms of similar descent and ecology, such as the fish, as a formal group, "Pisces." But in evolutionary terms, not all fish are created equal. Lungfish, for example, have a more recent common ancestor with four-legged land vertebrates (tetrapods) than they have with a shark or a tuna. In other words, a lungfish and a cow are more closely related than a lungfish and a tuna. Here we see a clear tension between ecological groupings, such as "fish," and evolutionary groups, such as the lungfish-tetrapod group (known as the Sarcopterygii). Which is better? The different priorities and goals of taxonomists led to much debate over the proper methods of classification. That debate still rages today. Some taxonomists argue that classification should be a matter of convenience, and they prefer using "Pisces" for fish, even though lungfish are really not closely related to other bony fish. Others insist that classification should reflect phylogeny, or evolutionary history, and nothing else. Thus the lungfish are put in a group with amphibians, not with other bony fish or sharks, because that is how they are related.

These changes can be jarring to those accustomed to the older traditional systems. People have often felt that birds are special, not only because many of us love watching and listening to them but also because they form a great diverse evolutionary radiation with many unique specializations including flight and feathers. These people wanted a class of birds called "Aves" to be equal in rank to the class Reptilia or the class Mammalia. But birds are descended from reptiles, specifically a group of predatory dinosaurs that include *Velociraptor*, so many taxonomists say that birds are just a subgroup of Dinosauria.

At one time, the family Pongidae was the formal grouping for the great apes (chimps, gorillas, orangutans, and gibbons), and there was another group of equal rank for humans (family Hominidae). Because of our anthropocentrism and arrogance, we humans insisted that we have our own family group that was separate from our ape relatives. But humans evolved from apes, and the Pongidae had no meaning except as an arbitrary clustering of the apes minus humans. Today the Pongidae has vanished from classifications, and the family Hominidae has been expanded to include all of our ape kin.

None of these arguments change the fact that all classifications show a branching, nested, hierachical pattern. The difference lies in whether our classification schemes should reflect this pattern strictly or should be a mixture of ecological factors (such as all swimming vertebrates being called "fish" or feathered vertebrates given their own class Aves) and phylogeny. Over the past 50 years, the trend has been to make classifications strictly a reflection of evolutionary history, and many traditional groups (such as Pisces) are gradually vanishing.

The classification schemes of Linnaeus and later taxonomists, which Darwin and others used, were based strictly on the visible anatomical features of animals and plants. Later, microscopic analysis of tissues and organ systems were added to the evidence for classification, refining them further. And the discovery of fossils has helped clarify the evolutionary past of different animals and plants and showed how they transformed through time. But neither Linnaeus nor Darwin nor any taxonomist before the 1960s realized that an additional level of information in organisms could test their hypotheses of relationships: molecular biology.

Think about it for a moment. Evolution predicts that anatomical features (both visible and microscopic and molecular) would show a nested pattern of similarity and produce a pattern of relationships suggesting the branching, bushy tree of life. Linnaeus documented this on the macroscopic level, and Darwin used powerful natural evidence to show that life had evolved. If life had been specially created rather than evolved, there would be no reason for the molecular systems to reflect this pattern of similarity seen in megascopic features. Molecular systems, for example, could have been created so that all animals that (say) live in a certain habitat have the same molecular patterns for developing the organs needed in that habitat. All the aquatic vertebrates, for example, from fish to penguins to whales and dolphins, could have identical molecular tool kits because they live in the water and need to do certain things, just as their external body form is streamlined for swimming and they all have flippers or fins for swimming as well.

But not even Darwin could have dreamed that the genetic code of every cell in your body also shows the evidence of evolution. Indeed, the pattern of molecular similarities places whales with other mammals (especially the hippopotamus), and penguins with the birds, and fish very far from either of them, just as the analysis of their internal organs and anatomy unrelated to swimming has always placed them. This first became apparent when the ability to detect certain molecules was developed with the invention of gel electrophoresis in the 1950s. It is one of the simplest, earliest, and least expensive techniques in molecular biology, and it has long been used to detect the presence of certain proteins. A concentrate of proteins is placed in a number of wells at the end of a thin sheet made of gel, and then an electrical field is applied across the gel. Different amino acids move at different rates in an electrical field (small molecules move faster than larger ones), so they "race" at different speeds across the gel as the field is applied. Once the field is turned off, the gel is stained, and the final position of each amino acid shows up as a dark band in its individual track.

Gel electophoresis gives a rough idea of which amino acids and proteins are present in a sample, although the method is not as precise for determining sequence as some other methods described here. Its main advantage is that it is relatively cheap, and it was very widely used in the 1960s and 1970s when it was the only technique available. It was once widely used to determine enzyme efficiency, genetic variability of natural populations, and gene flow and hybridization, and to recognize species boundaries and determine phylogenetic relationships.

Another popular method in the 1960s and 1970s was the measurement of immunological distance. Most animals produce antibodies that react to foreign substances (antigens) as part of their immune protection against disease and infection. When the antibodies from two organisms are mixed, the stronger the immunological reaction observed, the more similar two proteins are in their genetic sequence. Through a technique called micro-complement fixation, small amounts of antibodies of several different animals are placed in wells in a gel. The reaction between the antibodies in two different gels can be observed, and this reaction gives a semiquantitative measure of the degree of similarity. This method provides a crude approximation of the genetic similarity between two or more organisms, but was widely used in the 1960s and 1970s when no other methods were available. It, too, has largely been replaced today by direct DNA analysis.

In the 1960s, a better technique became available: amino acid sequencing. Widely used from the 1960s through the 1990s to determine the molecular similarities between many organisms (such as the hemoglobin, myoglobin, and cytochrome c), it was the first technique to produce branching sequences of relationships between organisms (figure 7.3). It remained popular as long as there was no means of identifying the actual DNA sequence that produced those amino acids and proteins.

The first method that directly measured the differences and similarities between DNA from two different kinds of organisms was DNA-DNA hybridization. If you boil a solution of DNA in water hotter than 100°C (212°F), the two strands of the molecule separate. As the solution cools, the individual strands seek to recombine with their exact matching strands. However,

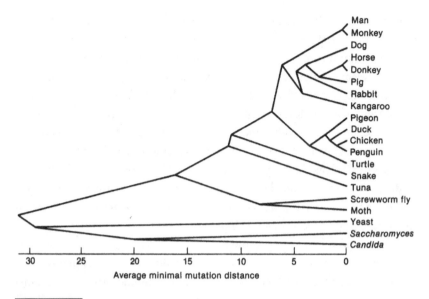

Figure 7.3 ▲

Branching diagram of the similarities in cytochrome c among various organisms. Nearly every biochemical system shows a similar branching pattern, which is identical to the branching pattern of life during its evolution. (From Walter M. Fitch and Emanuel Margoliash, "Construction of Phylogenetic Trees," *Science* 155, no. 3760 [January 1967]. Copyright © 1967 American Association for the Advancement of Science. Reprinted with permission.)

if you create a mixture of DNA from two different organisms, some strands from one organism will combine with the complementary strands from the other, forming a hybrid DNA. Then the mixture is reheated, breaking the DNA apart once again. The more similar the two strands are, the more tightly they will bond together, and a higher temperature is required to break them apart. DNA strands from two less closely related organisms, however, are less similar and less tightly bonded and will dissociate at lower temperatures.

DNA-DNA hybridization was extensively used in the 1970s and 1980s to record the percentage similarity between the DNA of two different organisms. It has revealed some startling things, such as the fact that humans and chimps share about 98 to 99 percent of their DNA (discussed in chapter 22). In 1983, DNA-DNA hybridization was used to completely reshuffle bird phylogeny and classification. However, it is not as informative as direct DNA sequencing because it provides only a semiquantitative estimate of similarity and not the actual base-by-base sequence.

Ultimately, the best method of determining molecular similarity is determining the direct sequence of the DNA itself. This has been the preferred method since the late 1990s, and it is at the cutting edge of molecular biology because it allows the scientist to directly determine each base pair along the strand of DNA (from the complete gene sequence of the nucleus to the independent genomes of organelles such as mitochondria and chloroplasts), so we can compare the gene sequence directly. Until the 1990s, it was difficult, slow, and too expensive to collect enough DNA to make this procedure possible. However, discovery of the polymerase chain reaction (PCR) amplification method revolutionized molecular genetics, making it possible to generate hundreds of copies of a DNA strand cheaply and quickly. Many of the new developments in genetics reported in the news come from DNA sequencing, and the multimillion-dollar Human Genome Project compiled the complete code for every gene in the human body in 2000 (simultaneously with Craig Venter's privately funded lab, Celera Genomics). In the past decades, hundreds of organisms have had their complete DNA sequenced, from bacteria and the nematode *Caenorhabditis elegans* to many different mammals to many primates, including all the great apes, and many humans as well.

Whether you look at the genetic sequence of mitochondrial DNA, nuclear DNA, cytochrome c, lens alpha crystallin, or any other biomolecule, the evidence is clear: The molecules show the same pattern of nested hierarchical similarity that the external anatomy reveals. Our molecules are most similar to those of our close relatives, the great apes, and progressively less similar to those more distantly related to us. We have found the proof of evolution in every cell in our bodies.

FOR FURTHER READING

Conniff, Richard. *The Species Seekers: Heroes, Fools, and the Mad Pursuit of Life on Earth*. New York: Norton, 2010.

Cracraft, Joel, and Michael J. Donoghue, eds. *Assembling the Tree of Life*. New York: Oxford University Press, 2004.

Eldredge, Niles, and Joel Cracraft. *Phylogenetic Patterns and the Evolutionary Process*. New York: Columbia University Press, 1980.

Felsenstein, Joseph. *Inferring Phylogenies*. Sunderland, Mass.: Sinauer, 2004.

Gould, Stephen Jay. "Linnaeus's Luck." *Natural History* 109, no. 7 (2000): 18–21.

Hillis, David M., and Craig Moritz, eds. *Molecular Systematics*. Sunderland, Mass.: Sinauer, 1990.

Lecointre, Guillaume, and Hervé Le Guyader. *The Tree of Life: A Phylogenetic Classification*. Trans. Karen McCoy. Cambridge, Mass.: Belknap Press of Harvard University Press, 2006.

Mayr, Ernst. *Principles of Systematic Zoology*. New York: McGraw-Hill, 1966.

Nei, Masatoshi, and Sudhir Kumar. *Molecular Evolution and Phylogenetics*. New York: Oxford University Press, 2000.

Page, Roderick D. M., and Edward C. Holmes. *Molecular Evolution: A Phylogenetic Approach*. Oxford: Blackwell Science, 1998.

Patterson, Colin, ed. *Molecules and Morphology in Evolution: Conflict or Compromise?* Cambridge: Cambridge University Press, 1987.

Simpson, George Gaylord. *Principles of Animal Taxonomy*. New York: Columbia University Press, 1961.

Yoon, Carol Kaesuk. *Naming Nature: The Clash Between Instinct and Science*. New York: Norton, 2010.

THE CASE OF THE CRUEL WASPS

What a book a devil's chaplain might write on the clumsy, wasteful, blundering, low, and horribly cruel work of nature!

—CHARLES DARWIN, IN AN 1856 LETTER TO JOSEPH HOOKER

In the late 1700s and early 1800s, the study of nature was not a science at all but was mostly a collection of casual observations about quaint and curious aspects of life. It was often called "natural philosophy" because it was based on philosophy and religion rather than on science (the concept of "science" would not emerge until the early 1800s). Much of the discussion was driven by those who were trying to "understand the mind of God through his handiworks." This branch of natural philosophy was called "natural theology," and most of the leading scholars were clergymen who had lots of time to study nature when not ministering to their flock. This was encouraged because they saw in nature the evidence of God's handiwork. Science was not yet a professional occupation.

The Reverend William Paley was the best known writer on this subject, and in 1802 he wrote *Natural Theology*, the most complete treatment of the subject. In that work, he described his famous "watchmaker" analogy. He asked the reader to imagine finding a watch and a rock on a beach. You would not consider the rock unusual because it is part of the natural world, but you would immediately recognize that the watch was "intricately contrived" and infer that it had been built by a watchmaker. To Paley, the intricate contrivances of nature were evidence that there was a Divine Watchmaker, namely, God.

The natural theology school of thought was very influential in its day, and Darwin himself knew Paley's book almost by heart. But natural theology had been debunked even before the time of Paley. In 1779, Scottish philosopher David Hume published *Dialogues Concerning Natural Religion*, which demolished the whole argument from design. Hume put the standard natural theology arguments in the mouth of a character called Cleanthes and used dialogues between him and a skeptic named Philo to tear down the design in nature argument. Philo notes that the design in nature analogy is faulty because we have no standard with which to compare our world, so it is possible to imagine a world much better designed than the one in which we live. Even if we concede that the world looks designed, it does not follow that the designer must be the Judaeo-Christian God. It could have been the god of another religion or culture, the work of a committee of gods, or a juvenile god who makes mistakes. Jews and Christians simply assumed that if there was a designer it must be their God, but there is no strong evidence to show that it wasn't some other god.

Nevertheless, the idea that nature was beautifully designed seemed compelling. Many books and poems have been written about the wonder of nature, but it is important for a scientist to look at the entire picture. As Darwin pointed out in the epigraph at the beginning of this chapter, nature can be "clumsy, wasteful, blundering, low and horribly cruel" as well. For centuries, writers and artists and religious folks looked at the beauty of the world without giving equal consideration and thought to its dark side. One of Darwin's great insights was that beauty and pain were equal parts of the story and could only be explained by a process that allowed both to operate. Nature is not just a divine display of the beautiful handiwork of a benevolent god; it is a process that operates outside our judgmental human framework of beautiful and ugly—it just gets a certain job done (survival of organisms so they can leave offspring to the next generation) by whatever means necessary. After all, we now know that life has been doing this for more than 3.5 billion years, and only in the last few thousand years have humans spent time rhapsodizing about the beauty of nature.

Biologists quickly learned to see both sides of this picture and to recognize that nature is more than pretty flowers or beautiful bird song. Life was not created simply to entertain us, and each living thing has a specific (and sometimes not so benevolent) function. Poets and painters marvel about gorgeous flowers (and we are still entranced by their appeal), but to a biologist, a flower is a sex organ. Its structure and function serve one main purpose—to move the male sperm in the pollen from one flower to reach the eggs in another flower, thereby maintaining a healthy, less inbred gene pool.

Every morning before dawn during the spring and summer, I hear mockingbirds singing on and on for hours with their incredible variety of birdcalls, some original to them, and some copied from other birds. To the naïve listener, the birds twittering in the trees are a lovely serenade for humans to enjoy. But a biologist recognizes that the function of singing these phrases over and over again is to issue a hostile warning to other birds of their species: "This my territory. Stay out!" This song is meant to drive other male mockingbirds away and to attract a female into his territory to mate with him. I get out my binoculars and spot one male mockingbird at the top of the tallest tree using all his energy all day long (and much of the night as well), with almost no time to hunt for food or rest his voice. I walk down the street a few blocks, and I hear a different male mockingbird at the top of another tall tree or telephone pole doing the same thing in his territory. These birds singing high above our heads are not singing for us—they are throwing down the gauntlet from their high perch, warning other male mockingbirds to keep the away from their turf. These insights and more are now among the first things you learn when you study biology, but in Darwin's day, these discoveries rocked the worldview of humans who believed flowers and bird song were created just for our enjoyment.

In 1829, the Earl of Bridgewater left a bequest of £8,000 to support a series of books "on the power, wisdom and goodness of God, as manifested in the creation." The first man to accept a commission to write one of the Bridgewater Treatises was the Rev. William Buckland at Oxford University. He was not only an Anglican cleric (and later Dean of Westminster) but also England's first official academic geologist. He named and published the first description of a dinosaur, *Megalosaurus*. Buckland loved animals so much that he and his family took pride in eating nearly every kind of animal they could obtain.

In his book *Geology and Mineralogy Considered with Reference to Natural Theology*, published in 1836, Buckland set about trying to give a theological explanation for "the problem of pain" in nature. Why would a benevolent deity let his creations suffer death and destruction? Why was there so much senseless cruelty in nature, whether it be a cat toying with a mouse before eating it or a large carnivore eating its prey while it is still alive and suffering? Buckland's answer to the problem was that carnivores actually increase "the aggregate animal enjoyment" and "diminish that of pain." If a prey animal dies swiftly when killed by a lion, the animal does not suffer the anguish of senility and disease and old age. Nor do the populations become so large that they exhaust their food supply and suffer the pangs of hunger and eventually a grim death of slow starvation. In Buckland's words,

> The appointment of death by the agency of carnivora as the ordinary termination of animal existence, appears therefore in its main results to be a dispensation of benevolence; it deducts much from the aggregate amount of the pain of universal death; it abridges, and almost annihilates, throughout the brute creation, the misery of disease, and accidental injuries, and lingering decay; and impose such salutary restraint upon excessive increase of numbers, that the supply of food maintains perpetually a due ratio to the demand. The result is, that the surface of the land and depths of the waters are ever crowded with myriads of animated beings, the pleasures of whose life are coextensive with its duration; and which throughout the little day of existence that is allotted to them, fulfill with joy the functions for which they were created.

Perhaps this view that predators were merely God's swift but merciful executioners gave some rationalization to the natural theology explanation of the problem of pain, but it certainly didn't help with the problem of parasites. People are naturally disgusted to learn that we all have thousands of bacteria, mites, and tiny roundworms living in various parts of our bodies; some benefit us, but many are not beneficial. We are even more revolted at the thought of huge tapeworms in our intestines, sapping our strength, or the many other parasites that can cripple, blind, or even kill us. This feeling of disgust made the original movie *Alien* a hit; seeing the parasitic alien creature burst out of the stomach of an astronaut (played by John Hurt) was truly nightmarish and horrible to most of us. Many natural theologians tried to rationalize the existence of parasites, but had no success. Buckland avoided the topic altogether because it completely undermined his whole argument about death in nature being a good thing.

No example puzzled and sickened the natural theologians more than the case of the "ichneumon fly" (actually an entire family of about 150,000 species of wasps, not flies, more species than all the vertebrates combined). Like most wasps, they live freely as adults, but they have a unique way of reproducing. When the female ichneumonid wasp is ready to lay her eggs, she finds a prey species, such as another insect or a spider, but most commonly a caterpillar. She swoops down on the unfortunate victim, pierces its body with the long "stinger" on her tail (actually an ovipositor, an organ for laying eggs), and injects venom that paralyzes the victim but leaves it alive (figure 8.1A). The mother lays her eggs inside the victim's body, and when the eggs hatch inside the caterpillar, the larvae begin to eat the host alive from the inside, starting with the less essential organs such as the digestive tract and fat bodies. The wasp larvae eat the nervous and circulatory systems last, which finally kills the host. This way their food is always alive and fresh until the very end, rather than dead and decaying before they finish their task. The caterpillar is now a hollow shell that becomes a protective case for the wasp larvae until they burst out of the victim's skin as fully flying adults. In some species, the female lays her eggs on top of the victim (figure 8.1B), and when the eggs hatch, the young wasps burrow inside the body and eat it alive.

The French entomologist Jean-Henri Fabre wrote whole books on this topic, describing many examples of parasitism by the larvae of wasps and other insects on their victims. In one species, their host may not be completely paralyzed, so the larvae are attached to a silken strand from the roof of the burrow and can retreat should their victim thrash around too much. As Fabre wrote in 1916 book, *The Hunting Wasps*:

> The grub is at dinner: head downwards, it is digging into the limp belly of one of the caterpillars. . . . At the least sign of danger in the heap of caterpillars, the larva retreats . . . and climbs back to the ceiling, where the swarming rabble cannot reach it. When peace is restored, it slides down [its silken cord] and returns to table, with its head over the viands and its rear upturned and ready to withdraw in case of need. (15)

Figure 8.1 ▲

Ichnemonid wasps: (*A*) A female wasp paralyzing and laying eggs inside an aphid. (*B*) These ichneumonid larvae parasitize the outside of caterpillars. (Courtesy of Wikimedia Commons)

Fabre also described wasps that parasitize crickets:

> One may see the cricket, bitten to the quick, vainly move its antennae and
> abdominal styles, open and close its empty jaws, and even move a foot, but
> the larva is safe and searches its vitals with impunity. What an awful night-
> mare for the paralyzed cricket! (16)

Fabre even did experiments in which he provided sugar water to the
partially consumed caterpillar, and it moved its mouthparts and attempted
to feed, showing that it was still alive and only partially paralyzed—even
though the wasp larvae had already eaten most of its insides, including its
digestive tract.

There are many other styles of parasitism as well. In 1982, Stephen Jay
Gould described a variety of styles of parasitism by wasps on different hosts:

> We learn of their skill in capturing dangerous hosts often many times larger
> than themselves. Caterpillars may be easy game, but psammocharid wasps
> prefer spiders. They must insert their ovipositors in a safe and precise spot.
> Some leave a paralyzed spider in its own burrow. *Planiceps hirsutus*, for exam-
> ple, parasitizes a California trapdoor spider. It searches for spider tubes on
> sand dunes, then digs into nearby sand to disturb the spider's home and drive
> it out. When the spider emerges, the wasp attacks, paralyzes its victim, drags
> it back into its own tube, shuts and fastens the trapdoor, and deposits a single
> egg upon the spider's abdomen. Other psamunocharids will drag a heavy spi-
> der back to a previously prepared cluster of clay or mud cells. Some amputate
> a spider's legs to make the passage easier. Others fly back over water, skim-
> ming a buoyant spider along the surface.
> Some wasps must battle with other parasites over a host's body. *Rhyssella
> curvipes* can detect the larvae of wood wasps deep within alder wood and drill
> down to a potential victim with its sharply ridged ovipositor. *Pseudorhyssa alpes-
> tris*, a related parasite, cannot drill directly into wood since its slender ovipositor
> bears only rudimentary cutting ridges. It locates the holes made by *Rhyssella*,
> inserts its ovipositor, and lays an egg on the host (already conveniently paraly-
> zed by *Rhyssella*), right next to the egg deposited by its relative. The two eggs
> hatch at about the same time, but the larva of *Pseudorhyssa* has a bigger head
> bearing much larger mandibles. *Pseudorhyssa* seizes the smaller *Rhyssella* larva,
> destroys it, and proceeds to feast upon a banquet already well prepared.

Other praises for the efficiency of mothers invoke the themes of early, quick, and often. Many ichneumons don't even wait for their hosts to develop into larvae, but parasitize the egg directly (larval wasps may then either drain the egg itself or enter the developing host larva). Others simply move fast. *Apanteles militaris* can deposit up to seventy-two eggs in a single second. Still others are doggedly persistent. *Aphidius gomezi* females produce up to 1,500 eggs and can parasitize as many as 600 aphids in a single working day. In a bizarre twist upon "often," some wasps indulge in polyembryony, a kind of iterated supertwinning. A single egg divides into cells that aggregate into as many as 500 individuals. Since some polyembryonic wasps parasitize caterpillars much larger than themselves and may lay up to six eggs in each, as many as 3,000 larvae may develop within, and feed upon a single host. These wasps are endoparasites and do not paralyze their victims. The caterpillars writhe back and forth, not (one suspects) from pain, but merely in response to the commotion induced by thousands of wasp larvae feeding within. (19–20)

Throughout the 1700s and 1800s, scholars and theologians wrestled with the horrific thoughts that were invoked by the stories of endoparasitism and tried to explain it away, or dismiss it, because it clashed so strongly with their notion that the universe was created and run by a benevolent deity. Most could not find a suitable explanation, however, because it never occurred to them that nature is not bound by human morality in the first place. Pioneering geologist Charles Lyell discussed ichneumonids in his landmark book, *Principles of Geology* (1830–1833). Even though wasps were far off topic for a book about geology, he rationalized that parasitic wasps were good for nature because otherwise caterpillars would destroy everything, especially human agriculture. In 1835, entomologist Rev. William Kirby wrote the seventh Bridgewater Treatise. He also considered caterpillars not worth saving, but he focused on the virtues of motherly love displayed by the wasps:

The great object of the female is to discover a proper nidus for her eggs. In search of this she is in constant motion. Is the caterpillar of a butterfly or moth the appropriate food for her young? You see her alight upon the plants where they are most usually to be met with, run quickly over them, carefully examining every leaf, and, having found the unfortunate object of her search, insert her sting into its flesh, and there deposit an egg. . . . The active Ichneumon braves every danger, and does not desist until her courage and address have insured subsistence for one of her future progeny.

Kirby even sympathized with the mother wasps, who never get to see their children alive:

> A very large proportion of them are doomed to die before their young come into existence. But in these the passion is not extinguished. . . . When you witness the solicitude with which they provide for the security and sustenance of their future young, you can scarcely deny to them love for a progeny they are never destined to behold.

And Kirby regarded the larvae as models of efficiency and a wise use of resources, eating their prey selectively so they stay fresh:

> In this strange and apparently cruel operation one circumstance is truly remarkable. The larva of the Ichneumon, though every day, perhaps for months, it gnaws the inside of the caterpillar, and though at last it has devoured almost every part of it except the skin and intestines, carefully all this time it avoids injuring the vital organs, as if aware that its own existence depends on that of the insect upon which it preys! . . . What would be the impression which a similar instance amongst the race of quadrupeds would make upon us? If, for example, an animal . . . should be found to feed upon the inside of a dog, devouring only those parts not essential to life, while it cautiously left uninjured the heart, arteries, lungs, and intestines,—should we not regard such an instance as a perfect prodigy, as an example of instinctive forbearance almost miraculous?

These passages from the early 1800s may strike us as odd, not only because of the religious assumptions about the benevolent deity but also because of their extremely anthropomorphic and anthropocentric tone. Not only do they use the language of human thoughts and feelings for insects who are driven by pure instinct, but they also assume that everything in nature is created for human benefit, one way or another. In reality, the ichneumonid reproductive system has probably been around for millions of years, possibly since the Permian (over 250 million years ago) when wasps first evolved, and certainly millions of years before humans were on the scene. Recent research on the family tree of the Hymenoptera (wasps, bees, and ants) shows that the ancestral hymenopteran was a wasp-like creature and that endoparasitism is a general feature of the whole order. Most of the groups of wasps alive today also have parasitic larvae. Only some of their descendants, including the more familiar vespid wasps, plus ants, bees, and termites, have lost this mode of reproduction.

Most important, it demonstrates the fallacy of reading moral meaning from nature. Nature is what it is—its glories and horrors are not somehow guides for our own moral decisions. Darwin himself was troubled by it, as he wrote in a letter to Asa Gray in 1860:

> I own that I cannot see as plainly as others do, and as I should wish to do, evidence of design and beneficence on all sides of us. There seems to me too much misery in the world. I cannot persuade myself that a beneficent and omnipotent God would have designedly created the Ichneumonidae with the express intention of their feeding within the living bodies of Caterpillars, or that a cat should play with mice.

Nature is neither good nor bad—it is just as we find it. We cannot learn moral lessons from, nor should we impose our own morality on, a nonmoral world. Trying to frame the meaning of nature "in our terms" is inappropriate because nature was not made for us. Nature is ruled by the impersonal laws of physics and chemistry and biology, including natural selection. These laws of nature do not care about pain or cruelty or joy or beauty; nature's laws are directed toward organisms successfully leaving descendants in the next generation.

Biologist Julian Huxley, grandson of Darwin's strongest advocate, Thomas Henry Huxley, put it this way in *Evolution: The Modern Synthesis*, published in 1943:

> Natural selection, in fact, though like the mills of God in grinding slowly and grinding small, has few other attributes that a civilized religion would call divine. . . . Its products are just as likely to be aesthetically, morally, or intellectually repulsive to us as they are to be attractive. We need only think of the ugliness of Sacculina or a bladder-worm, the stupidity of a rhinoceros or a stegosaur, the horror of a female mantis devouring its mate or a brood of ichneumon flies slowly eating out a caterpillar. (485)

Or as Stephen Jay Gould wrote in "Nonmoral Nature" in 1982:

[The] natural world [is] neither made for us nor ruled by us. It just plain happens. It is a strategy that works for ichneumons and that natural selection has programmed into their behavioral repertoire. Caterpillars are not suffering to teach us something; they have simply been outmaneuvered, for now, in the evolutionary game. Perhaps they will evolve a set of adequate defenses sometime in the future, thus sealing the fate of ichneumons. And perhaps, indeed probably, they will not. (21)

FOR FURTHER READING

Branstetter, Michael G., Bryan N. Danforth, James P. Pitts, Brant C. Faircloth, Philip S. Ward, Matthew L. Buffington, Michael W. Gates, Robert R. Kula, and Seán G. Brady. "Phylogenomic Insights into the Evolution of Stinging Wasps and the Origins of Ants and Bees." *Current Biology* 27, no. 7 (2017): 1019–1025.

Fabre, Jean-Henri. *The Hunting Wasps*. London: Hodder and Stoughton, 1916.

Gould, Stephen Jay. "Nonmoral Nature." *Natural History* 91 (1982): 19–26

Kirby, William. *On the Power Wisdom and Goodness of God. As Manifested in the Creation of Animals and in Their History, Habits and Instincts*. London: W. Pickering, 1835.

Peters, Ralph S., Lars Krogmann, Christoph Mayer, Alexander Donath, Simon Gunkel, Karen Meusemann, Alexey Kozlov, Lars Podsiadlowski, Malte Petersen, Robert Lanfear, Patricia A. Diez, John Heraty, Karl M. Kjer, Seraina Klopfstein, Rudolf Meier, Carlo Polidori, Thomas Schmitt, Shanlin Liu, Xin Zhou, Torsten Wappler, Jes Rust, Bernhard Misof, and Oliver Niehuis. "Evolutionary History of the Hymenoptera." *Current Biology* 27, no. 7 (2017): 1013–1018.

JURY-RIGGED CONTRIVANCES

It is only by the display of contrivance, that the existence, the agency, the wisdom of the Deity, could be testified to his rational creatures. This is the scale by which we ascend to all the knowledge of our Creator which we possess, so far as it depends upon the phenomena, or the works of nature . . . it is in the construction of instruments, and the choice and adaptation of means, that a creative intelligence is seen. It is this which constitutes the order and the beauty of the universe.

—WILLIAM PALEY, *NATURAL THEOLOGY* (1802)

I do not think I hardly ever admired a work more than Paley's "Natural Theology." I could almost formally have said it by heart.

—CHARLES DARWIN, IN AN 1887 LETTER TO HIS NEIGHBOR JOHN LUBBOCK

From Robert Boyle's 1688 *Disquisition About the Final Causes of Natural Things* and John Ray's 1691 *Wisdom of God Manifested in the Works of Creation*, there was a century-old tradition in the devout parts of Europe to point to the beauty and harmony and intricate "design" of nature as evidence of God's handiwork. Indeed, many natural historians were also clergymen; they saw studying natural history as a way to better understand the mind of the Creator. As discussed in chapter 8, the most famous advocate of natural theology was the Rev. William Paley, who in 1802 wrote *Natural Theology*, the classic treatment of the subject.

As Darwin's ideas evolved when he returned from his voyage around the world on the *Beagle*, he naturally recalled his reading of Paley. At the same time, however, his close study of natural history (especially the barnacles, which he studied until he had become the world's expert on them) showed him the perfect counterargument to Paley. The beauty and intricacy of nature was evident in many places, but so was the evidence of "contrivances" that were not perfect or optimal but "jury-rigged" to work just well enough for the organism to survive long enough to leave offspring in the next generation. As Darwin and later biologists came to realize, nature does not require perfection. It is about adaptation to local circumstances, and the organism's design does not have to be perfect to ensure survival. Moreover, once an organism successfully reproduces, there is no longer any selection pressure on them (unless the parents are required to raise the offspring to the next generation). Many species (most invertebrates and fish) mate, lay their eggs, and then die right away—and that is all that natural selection requires. A suboptimal jury-rigged solution to life's problems is good enough as long as the organism can successfully reproduce.

Darwin focused on these suboptimal, jury-rigged contrivances in nature that seemed to suggest a clumsy or sloppy designer but are consistent with the idea that nature does not require perfection. A solution that works well enough for survival and reproduction is sufficient. Darwin himself realized how important it was to illustrate his argument, so his next book after *On the Origin of Species* in 1859 was not on the controversial topic of human evolution (he didn't touch that subject until 1871) but on the cross-fertilization of orchids.

Why follow the most important book and idea in the history of biology with a book on orchids? As Darwin cultivated and studied orchids in his greenhouse in his backyard (figure 9.1), he found that their flowers were full of jury-rigged apparatuses that were not perfect but just good enough to attract insects and promote cross-fertilization. He also corresponded with many other orchid growers across the world and got to know them very well. In 1862 in *On the Various Contrivances by Which British and Foreign Orchids Are Fertilized by Insects*, Darwin wrote:

> Although an organ may not have been originally formed for some special purpose, if it now serves for this end we are justified in saying that it is specially contrived for it. On the same principle, if a man were to make a machine for some special purpose, but were to use old wheels, springs, and pulleys, only slightly altered, the whole machine, with all its parts, might be said to

Figure 9.1 ▲

Darwin wrote a whole book about the jury-rigged devices that orchid flowers use to ensure fertilization whenever a vector such as a bee visits them. He based this book on his observations of cultivating many orchids in the backyard greenhouse at his estate, Down House. This is a shot inside Darwin's own greenhouse of some orchids still being grown there. (Photograph by the author)

be specially contrived for that purpose. Thus throughout nature almost every part of each living being has probably served, in a slightly modified condition, for diverse purposes, and has acted in the living machinery of many ancient and distinct specific forms.

As Stephen Jay Gould wrote in a *New Scientist* essay titled "The Panda's Thumb and the Orchid's Trap" in November 1978:

Orchids have formed an alliance with insects. They have evolved an astonishing variety of "contrivances" to attract insects, guarantee that their sticky pollen adheres to the visitor, and ensure that the attached pollen comes in contact with female parts of the next orchid the insect visits.

Darwin's book is a compendium of these contrivances, the botanical equivalent of a bestiary. And like the medieval bestiaries, it is designed to instruct. The message is paradoxical but profound. Orchids manufacture their intricate devices from the common components of flowers, parts usually fitted for very different functions. If God had designed a beautiful machine to reflect his wisdom and power, surely he would not have used a collection of parts generally fashioned for other purposes. Orchids were not made by an ideal engineer; they are jury-rigged from a limited set of available components. Thus, they must have evolved from ordinary flowers.

Thus, the paradox: Our textbooks like to illustrate evolution with examples of optimal design—nearly perfect mimicry of a dead leaf by a butterfly or of a poisonous species by a palatable relative. But ideal design is a lousy argument for evolution, for it mimics the postulated action of an omnipotent creator. Odd arrangements and funny solutions are the proof of evolution—paths that a sensible God would never tread but that a natural process, constrained by history, follows perforce. No one understood this better than Darwin. (700)

An entire book on orchids formed Darwin's best example of jury-rigged organisms that have modified whatever anatomy they had available for totally new uses. But the examples can be multiplied endlessly. Three examples are well known because they have been discussed many times, especially by Stephen Jay Gould.

One of the weirdest examples is a freshwater clam called *Lampsilis* (figure 9.2A). Like most freshwater clams, it spends most of its life buried in the sand of a creek, lake, or river, filtering out food from the water passing over its gills. But like certain freshwater clams and snails, it spreads its larvae to other places by having them latch onto the gills of a fish, where they complete their growth. *Lampsilis* accomplishes this macabre way of spreading its larvae in a crude but effective manner. The brood pouch full of eggs protrudes from the rear of the clam's shell, where it has a shape that vaguely resembles a fish. There is a crude "eye spot" and a fringe that ripples in the current like a fish's fin. Once a curious fish gets close enough to bite it, the clam ejects the larvae, which are swallowed by the fish and hook onto its gills to complete their development. The "fishing lure" is not even a good imitation of a fish—but it doesn't have to be because fish don't have great eyesight and movement is often a more important cue. Anyone who has crafted fishing lures knows this—they don't have to match a real animal that well. As long as the lure looks reasonably like a prey item and moves in the right way, it will attract the fish.

Another weird but effective example of a crude fishing lure is employed by the anglerfish (figure 9.2B; figure 9.2C). This ugly brute looks like a bumpy rock on the seafloor, and it rarely moves very far. Instead, it ambushes its prey by luring them closer until it can suck them down its throat with one big gulp. How does it entice its prey close enough to ambush them? Above its mouth is a long, highly modified spine that it waves back and forth in the water. On the tip of the spine is a fringe that has the vague shape of a fish. It's not even close to resembling any real fish—just a fusiform shape with a few dark stripes—but it's moving back and forth like a fish, and that's all it takes to get curious prey fish to swim close enough to be gulped down.

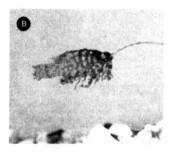

Figure 9.2 ▲

Nature is full of examples of jury-rigged adaptations that work just well enough to serve a purpose even though they are not perfectly designed. (A) The freshwater clam *Lampsilis* has a brood pouch that looks somewhat like a fish and lures fish to bite it. When they do, the clam's larvae hook onto the fish's gills and complete their life cycle. (B) The anglerfish has a spine above its mouth with a fringed tip that looks vaguely fish-like. (C) When prey comes near to bite the lure, the anglerfish sucks its victim into its mouth. ([A] Photograph by J. H. Welsh, from the cover of *Science* 134, no. 3472 [1969]. Copyright © 1969 American Association for the Advancement of Science. Reprinted with permission; [B–C] photographs from Theodore W. Pietsch and David B. Grobecker, *Science* 201, no. 4353 [1978]: 369–370. Copyright © 1978 American Association for the Advancement of Science. Reprinted with permission.)

One of the most famous examples of jury-rigged, half-assed "designs" was Stephen Jay Gould's favorite case, the panda's thumb (figure 9.3). The giant panda (*Ailuropoda melanoleuca*) is a member of the order Carnivora, distantly related to all other carnivorans (dogs, cats, weasels, skunks, hyenas, raccoons, and their kin), and among them it is most closely related to bears. Indeed, it has often been called the "panda bear." Like all other

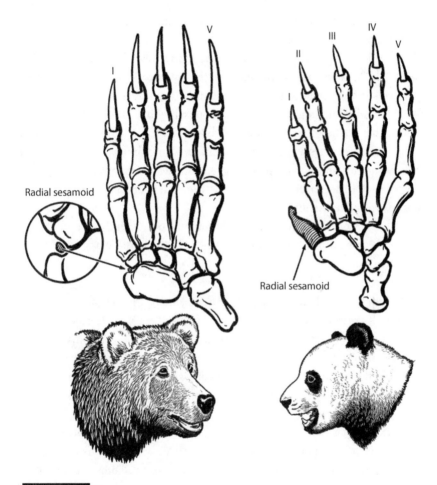

Figure 9.3 ▲

The panda, like all Carnivora, has all five fingers forming a paw, but unlike other Carnivora, it eats bamboo. Consequently, it has modified a wrist bone, the radial sesamoid, into a crude "thumb" that enables it to strip the leaves off bamboo. It works just well enough to feed a panda; it is not beautifully designed, but crude, clumsy, and jury-rigged. (Drawing by Carl Buell)

carnivorans, its true thumb is part of its paw and is tightly pressed to the side of the paw, so it has little use (like the "dew claw" in a cat's or dog's paw). But unlike the rest of the Carnivora, pandas eat plants (bamboo), not meat, so their lifestyle is very different.

Most of the time pandas live in the high mountains of China. They are too big for most predators to threaten, and they eat hundreds of pounds of bamboo stalks every day, 12 to 14 hours a day. They hold the stalks of bamboo in their paws and strip off the leaves by running the stalk between their paw and what appears to be a thumb. Their true thumb is tiny and fused to the paw, so they have jury-rigged a bone in the wrist, known as the radial sesamoid, to serve as a "sixth finger." It is a patchwork "thumb" that serves just well enough to strip leaves from bamboo stalks. It's nowhere near as flexible and strong and useful as the opposable thumb of primates, but it doesn't have to be—the panda just needs a device for stripping leaves. These examples demonstrate that the key to nature is not in its perfection but in how temporary solutions and jury-rigged contrivances show that organisms do not need to be perfectly engineered—they just need to work well enough to survive and leave offspring.

In a 1978 essay, "The Panda's Thumb and the Orchid's Trap," Gould wrote about what the anatomist D. Dwight Davis found when he studied the giant panda:

> As with the radial sesamoid, these muscles did not arise *de novo*; like the parts of Darwin's orchids, they are familiar bits of anatomy, remodeled for a new function. The abductor of the radial sesamoid (the muscle that pulls it away from the true digits) bears the formidable name *M. abductor pollicis longus* (the long abductor of the thumb—pollicis is the genitive of pollex, Latin for thumb). Its name is a giveaway. In other carnivores, it attaches to the first digit, or true thumb. . . .
>
> Does the anatomy of other carnivores give us any clue to the origin of this odd arrangement in pandas? Davis points out that ordinary bears and raccoons, the closest relatives of giant pandas, far surpass all other carnivores in using their forelegs for manipulating objects in feeding. Pardon the backward metaphor, but pandas, thanks to their ancestry, began with a leg up for evolving greater dexterity in feeding. Moreover, ordinary bears already have a slightly enlarged radial sesamoid. In most carnivores, the same muscles that move the radial sesamoid in pandas attach exclusively to the base of the

pollex, or true thumb. But in ordinary bears, the long abductor muscle ends in two tendons: one inserts into the base of the thumb as in most carnivores, but the other attaches to the radial sesamoid. Two shorter muscles . . . also attach, in part, to the radial sesamoid in bears. "Thus," Davis concludes, "the musculature for operating this remarkable new mechanism—functionally a new digit—required no intrinsic change from conditions already present in the panda's closest relatives, the bears. Furthermore, it appears that the whole sequence of events in the musculature follows automatically from simple hypertrophy of the sesamoid bone." (700–701)

The panda's thumb has been a popular example for a long time, but recent research has given the story yet another twist. Along with the giant panda, there is another bamboo-eating carnivoran in eastern Asia, the "lesser panda" or "red panda" (*Ailurus fulgens*). It is a reddish brown color and is about the size and shape of a cat with a bushy tail and red and white stripes on its tail and body (figure 9.4). Its face has a mask of black stripes

Figure 9.4 ▲

Photographs of the lesser, or "red," panda, showing its extraordinarily broad forepaws with a "thumb" that allows it to grip tree bark or grasp leaves and stems. (Courtesy of Wikimedia Commons)

and a white muzzle, and its ears are black and white as well. It also lives in the highest mountains in China, Tibet, and Nepal, coping with snowy winters and eating other kinds of small prey when bamboo is scarce.

For many years, it was thought to be a very close relative of the giant panda because it also has the same weird "thumb" jury-rigged from the radial sesamoid of its wrist and used it to strip leaves from bamboo, its main food source. But in 2000, a molecular study was made of its DNA, and it turns out not to be related to giant pandas at all; rather, it belongs to the carnivoran group Musteloidea, along with skunks, weasels, and raccoons. This is consistent with lots of earlier research in their karyology, serology, behavior, anatomy, and reproduction, and especially their fossil record, which allies them more closely with raccoons and weasels, not bears and pandas.

If this is true (and the evidence is overwhelming now), then the weird jury-rigged panda's thumb has twice evolved independently in two unrelated groups of carnivorans. In earlier days, scientists might have regarded such a specialized paw structure as a truly unique feature and could not have imagined that it could evolve in parallel. But fossil, molecular, and behavioral data cannot be denied, so we are forced to admit that it did so. Keep in mind, however, that both animals came from common ancestors (like bears and raccoons) that already had highly flexible grasping forepaws with a small radial sesamoid partially developed. And they both needed a solution to stripping leaves from bamboo stalks and have the same basic tool kit of a paw with a tiny "dew claw" thumb that cannot be made opposable. They have both started with the same basic anatomy and modified it for a similar function, which led to a similar solution.

Examples of jury-rigged, suboptimally designed features occur throughout nature, especially in humans (see chapter 21), but these few examples make the main point. If nature had been created by a perfect Divine Designer who used the best, most elegant solutions and engineered everything to work perfectly and efficiently, we would not see so many examples of organism that use clumsy, suboptimal, jury-rigged anatomy to survive just long enough to breed.

FOR FURTHER READING

Davis, D. Dwight. *The Giant Panda: A Morphological Study of an Evolutionary Mechanism.* Chicago: Chicago Natural History Museum, 1964.

Flynn, John J., Michael A. Nedbal, Jerry W. Dragoo, and Rodney L. Honeycutt. "Whence the Red Panda?" *Molecular Phylogenetics and Evolution* 17, no. 2 (2000): 190–199.

Gould, Stephen Jay. *The Panda's Thumb*. New York: Norton, 1980.

Glatston, Angela R. *Red Panda: Biology and Conservation of the First Panda*. Amsterdam: Elsevier, 2011.

Pietsch, Theodore W., and David B. Groebecker. "The Compleat Angler: Aggressive Mimicry in the Antenariid Anglerfish." *Science* 201, no. 4353 (1978): 369–370.

Slattery, J. Pecon, and S. J. O'Brien. "Molecular Phylogeny of the Red Panda (*Ailurus fulgens*)." *Journal of Heredity* 86, no. 6 (1995): 413–422.

PART III

GREAT TRANSITIONS IN THE HISTORY OF LIFE

A WHALE OF A TALE

Rudimentary, atrophied, or aborted organs.—Organs or parts in this strange condition, bearing the stamp of inutility, are extremely common throughout nature. For instance, rudimentary mammæ are very general in the males of mammals: I presume that the "bastard-wing" in birds may be safely considered as a digit in a rudimentary state; in very many snakes one lobe of the lungs is rudimentary; in other snakes there are rudiments of the pelvis and hind limbs. Some of the cases of rudimentary organs are extremely curious; for instance, the presence of teeth in fœtal whales, which when grown up have not a tooth in their heads; and the presence of teeth, which never cut through the gums, in the upper jaws of our unborn calves. It has even been stated on good authority that rudiments of teeth can be detected in the beaks of certain embryonic birds. Nothing can be plainer than that wings are formed for flight, yet in how many insects do we see wings so reduced in size as to be utterly incapable of flight, and not rarely lying under wing-cases, firmly soldered together!

—CHARLES DARWIN, *ON THE ORIGIN OF SPECIES* (1859)

Among Darwin's most powerful evidence for the fact of evolution were the many examples of organs in living animals that were shrunken, useless, or otherwise no longer fully functional (in Darwin's words, "rudimentary, atrophied, or aborted"). These had been noted by naturalists before him, who struggled to explain why a Divine Designer would include such worthless structures in organisms that were clearly not using them. These theologian-naturalists tried out many inadequate explanations—the organs were there to maintain the symmetry of the design; their presence showed us that the designer could do whatever he wanted; or flaws in design were due to Adam's failure to obey God in the Garden of Eden—but most of them were an embarrassment and left unmentioned.

For Darwin, however, the answer was clear: These organs were silent witnesses to the fact that these animals had stopped needing these once functioning organs. The organs had shrunk and become useless, but they had not vanished. If animals had been independently created and designed from scratch, why include organ systems that were useless or inefficient? Why not design the organism to have maximum efficiency with no wasted parts? Clearly, they made no sense if an organism had been divinely designed. This seemed to imply that the designer was lazy or careless or less than competent because he did not remove organs that had no use. Just as in the discussion in chapter 9, if there had been a Divine Designer, surely he would have done a better job!

The list of strange but useless features is quite striking (figure 10.1):

1. Horses have tiny splints on their side toes, remnants of the days when all horses had three functioning toes (chapter 14).

2. Boas and pythons, and some other snakes, have remnants of hip bones and thighbones deeply embedded in their bodies that perform no function (chapter 12).

3. Numerous fish, salamanders, crickets, and other animals live in caves in total darkness all of their lives and are blind, yet they have eyes that develop like normal eyes. If they had been divinely created to live in total darkness, why bother to develop useless eyes? Experiments show that the same strain of fish, raised in a well-lit setting for a few generations, will regain the ability to see.

4. Wings on flightless birds make no sense unless their ancestors were once birds with flight. Most of these birds are also too large to fly, so the wing is even more pointless. They don't use the wings for anything, yet they go the trouble of developing them. This has happened not only in the ratites (ostriches, emus, cassowaries, rheas, kiwis) but also in many other groups (such as dodos). Flightlessness is especially common on islands where birds no longer need flight to escape predators. Darwin himself commented on how the cormorants of the Galápagos Islands were flightless and had tiny, stunted wings (figure 10.2). They were perfectly able to dive off cliffs and fish as do other cormorants, but they didn't need to fly to reach the water or to escape predators.

5. Human beings have not escaped this trend. Hundreds of examples of useless features are included in the human body (chapter 21).

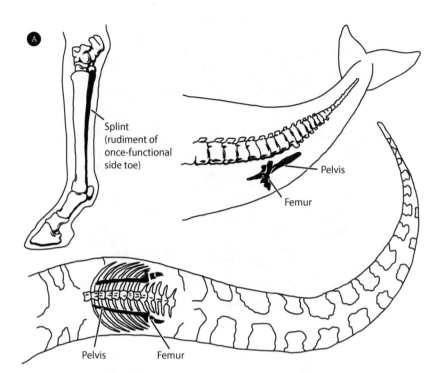

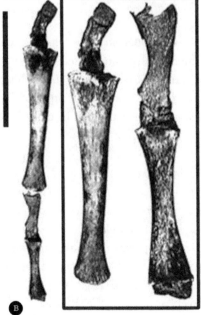

One foot

Figure 10.1 ▲

The evidence from vestigial organs. (A) Both whales and snakes retain tiny remnants of their hind legs and hip bones, although they are normally not externally visible, nor do they have any function. These organs only make sense if whales and snakes were once four-legged creatures. Horses also retain vestiges of their ancestral side toes, known as splint bones. (B) In 1921, Roy Chapman Andrews documented a specimen of a humpback whale that had atavistic hind limbs that actually extended from its body. These are the bones of those hind limbs. ([A] From Donald Prothero, *Evolution: What the Fossils Say and Why It Matters*, 2nd ed. [New York: Columbia University Press, 2017]); [B] Roy Chapman Andrews, "A Remarkable Case of External Hind Limbs in a Humpback Whale," *American Museum Novitates*, no. 9 [1921], http://hdl .handle.net/2246/4849)

Figure 10.2 ▲

Charles Darwin described cormorants from the Galápagos Islands that had stunted, stubby wings and were unable to fly—since no predators from which they needed to flee lived on those islands. Their ability to dive and swim underwater was not hampered by the reduced size of their wings. (Courtesy of Wikimedia Commons)

In addition, dozens of examples of molecular structures have been described that are unnecessary or inefficiently constructed or poorly designed. As Behrman, Marzluf, and Bentley commented, "In teaching metabolic pathways, every instructor emphasizes the chemical logic of the transformations wherever possible. In cases such as those to be described here, the lecturer is reduced to impotent hand waving."

1. Unnecessary transformation of a carbon atom from "left-handed" (S) to "right-handed" (R) chirality in the propionyl CoA → succinyl CoA pathway.
2. The pathways for several compounds begin by converting (S)-reticuline to (R)-reticuline. It would be simpler to just use S-reticuline.
3. During the biosynthesis of the amino acid tryptophan, a 3-carbon fragment is removed (glyceraldehyde 3-phosphate), and in the next step, a 3-carbon fragment (serine) is added back (actually, in this example, it seems that adding serine is probably a more direct way of getting the "backbone" part of the amino acid)—it would take three steps to convert the glyceraldehyde

3-phosphate chain (-CHOH-CHOH-CHOPO$_3$) into the amino acid carbon chain (CHH-CHNH$_3$-COO).

4. When DNA strands are separated and being replicated, the "leading strand" is continuously copied by a $5' \rightarrow 3'$ DNA polymerase. The "lagging strand" is unwound $3' \rightarrow 5'$, but because replication proceeds in the $5' \rightarrow 3'$ direction on the DNA strand, the lagging strand must be synthesized in short bits that are then pieced together. It is pointed out that it would be simpler and more logical just to use a $3'-5'$ DNA polymerase on the lagging strand, but this is not what biology does.

5. Unnecessary RNA editing. A pre-RNA for an ion channel protein has a particular codon, CAG, that codes for glutamine. However, if the protein that is produced from the RNA contains that glutamine, the mouse dies. In healthy mice, another protein edits the pre-RNA and changes CAG to CIG, which is equivalent to CGG and produces arginine, the correct amino acid. It would be much simpler for the pre-RNA to just code for CGG to start with, and dispense with the editing step. Indeed, scientists have made this change in the lab, and the resulting mice are normal and no longer need the editing step.[1]

This list of unneeded features could go on and on; there are literally hundreds of examples. But let's focus on one group in particular, the whales. They have many structures that are vestigial or poorly designed: the tiny vestigial hip bones and thighbones deeply embedded in their bodies with no real function (figure 10.1A; figure 10.3), the vestigial teeth of otherwise toothless baleen whales, and many other features.

Think about the clumsiness of this process. A whale embryo has gill slits for breathing water, but later in embryology it loses them and develops lungs. After the whale is born, it must compensate and develop many unusual structures to make it possible to breathe air while being fully aquatic. If whales had been designed for life in the water, why allow their embryos to lose their perfectly good gills and replace them with a jury-rigged system of lungs and other strange adaptations to enable an air-breathing lung to function in a marine mammal? Whales have all sorts of body modifications to make it easier to breathe in the ocean, such as moving the nasal opening from the tip of the snout to the top of the skull, where it acts as a blowhole. But aquatic life is challenging in many ways if you're not a fish. In addition to the problem of breathing with lungs, whales give live birth underwater,

Figure 10.3 ▲

Close-up of the hip region of a mounted blue whale skeleton, showing the tiny vestigial hip bones and hanging from the triangular bracket below the spine. (Photograph by the author)

and the whale calf must find a way to the surface to get its first breath. Observations from the field have shown that this is a very dangerous stage; if the newborn whale fails to do this quickly, it will drown.

Why do whales have these strange adaptations? The answer, of course, is that whales are descended from four-legged land mammals that once breathed air on land and were not aquatic. Even in the 1750s, the father of modern classification, Carolus Linnaeus, realized that whales were not fish but mammals (most of his contemporaries classified them as fish). As he pointed out, whales breathed air through lungs, not gills, and were warm-blooded, and had many other anatomical differences that distinguished them from fish. In the first edition of *On the Origin of Species* in 1859, Darwin also weighed in on this topic. He wrote: "In North America the black bear was seen by Hearne swimming for hours with widely open mouth, thus catching, like a whale, insects in the water. Even in so extreme a case as this, if the supply of insects were constant, and if better adapted competitors did not already exist in the country, I can see no difficulty in a race of bears being rendered, by natural selection, more aquatic in their structure and habits, with larger and larger mouths, till a creature was produced as monstrous as a whale." Unfortunately, this idea didn't go over so well with Darwin's critics, and he dropped this speculation from some of the later editions of his book.

And that's where the question stood, unanswered for about a century. There were a few skeletons of very primitive whales called archaeocetes from the Eocene beds of the southeastern United States, but no fossils that showed how whales evolved from terrestrial mammals, nor any indication of who their closest relatives might be among the mammals. In 1966, paleontologist Leigh Van Valen reopened the question, pointing out that an extinct group of bear-like carnivorous hoofed mammals known as mesonychids had large blunt triangular teeth and other features very much like the archaeocete whales. Whales even had huge skulls with long snouts, very similar to some mesonychids. Bit by bit, the idea that whales evolved from mesonychids became more acceptable.

Meanwhile, several other groups of scientists were searching the Eocene rocks of Pakistan, looking for fossils from the early middle Eocene, slightly older than the archaeocete whales of Alabama and Egypt. Sure enough, they began to find more and more fossils that were transitional between land animals and whales.

The oldest transitional whale fossil very close to the early anthracotheres is *Indohyus*, a collie-sized fossil from the early Eocene of Kashmir. Even though it was barely larger than a rabbit, with long hind legs for leaping and the body of a small deer, it had distinctive anatomical features that make it the link between whales and other artiodactyls. Its ear region shows many features that are found only in the whales. It also has limbs made of very dense bone (just like whales, hippopotamuses, and many other aquatic groups) that provide ballast and help it wade or dive underwater without floating out of control. Chemical analysis of the bones showed they were aquatic, but the chemistry of their teeth proved that they ate land plants.

In 1983, Philip Gingerich and his colleagues found a fossil they called *Pakicetus*, named for the country where it was found. It was a mostly terrestrial animal, with a wolf-like skeleton with four walking legs, but its skull was like that of an archaeocete whale, and there were features of the ear region that might help with underwater hearing. It came from river sediments that dated to about 50 million years ago, older than any archaeocete. Even though its limbs were wolf-like for running, they were made of very dense bone common in aquatic animals, suggesting that it spent a lot of time swimming in the water.

In 1994, Hans Thewissen and colleagues found and reported a true transitional fossil (figure 10.4) from beds in Pakistan that were about 47 million years old, just slightly younger than *Pakicetus*. It was from the Kuldana Formation, which has not only nearshore marine beds but also lake and river deposits. They named it *Ambulocetus natans*, which translates as "walking swimming whale." It was a nearly complete skeleton about 3 meters (10 feet) long, about as big as a large sea lion. Its long snout with triangular teeth was much like mesonychids and archaeocete whales, but it had fully developed arms and legs with long fingers and probably webbed feet and was capable of walking both on land and swimming in the water. Its spine was flexible, so it could swim in an up-and-down undulation as an otter does rather than having a rigid body and using its propulsive tail like a modern whale. But it was clearly not a fast swimmer. Its long snout and body and large feet with webbed toes were more like that of a crocodile, suggesting that it was more of an ambush predator, lurking semisubmerged in shallow water and lunging out of the water to grab prey that came to drink (figure 10.4B). Chemical analysis of the bones showed it was primarily aquatic but lived in both fresh and salt water. As Stephen Jay Gould wrote when this amazing transitional fossil that defied the critics of evolution was found:

> These dogmatists, who by verbal trickery can make white black, and black white, will never be convinced of anything, but *Ambulocetus* is the very animal that they proclaimed impossible in theory. . . . I cannot imagine a better tale for popular presentation of science or a more satisfying, and intellectually based political victory over lingering creationist opposition. (10)

The same year that *Ambulocetus* was found in one part of Pakistan, Gingerich and colleagues discovered and reported another fossil they named *Rodhocetus* from different beds in the Balochistan region of Pakistan (figure 10.5). Its skull is even longer and more whale-like than *Ambulocetus*, and the body is quite dolphin-like with a streamlined shape and no neck to separate the body from the head. More to the point, its limbs are very short

Figure 10.4 ▲

Ambulocetus natans, the primitive whale from the Eocene of Pakistan that still retains a mesonychid-like head, large functional webbed hands and feet, and a semiaquatic mode of life. (*A*) A nearly complete skeleton laid out in anatomical position. (*B*) A reconstruction of *Ambulocetus* lunging out of the water to capture another Eocene mammal. ([*A*] Courtesy of J. G. M. Thewissen; [*B*] reconstruction by Carl Buell)

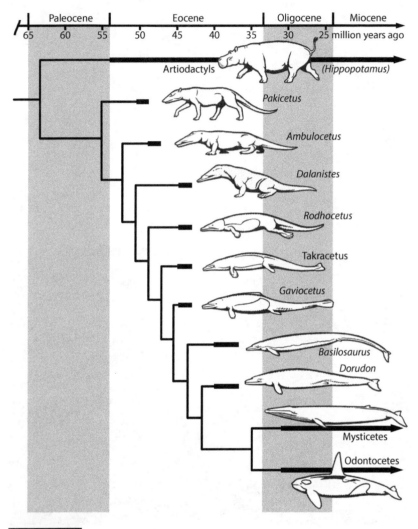

Artiodactyls *(Hippopotamus)*

Pakicetus

Ambulocetus

Dalanistes

Rodhocetus

Takracetus

Gaviocetus

Basilosaurus
Dorudon

Mysticetes

Odontocetes

Figure 10.5 ▲

Evolution of whales from land creatures, showing the numerous transitional fossils now documented from the Eocene beds of Africa and Pakistan. (Drawing by Carl Buell)

and small with webbed feet, not very effective for dragging its body across land, so it was primarily aquatic. But it did have hip bones and hip vertebrae fused together, so it was capable of limited land locomotion. The structure of the limbs and tail, however, suggest that it mostly stroked with its hind feet for propulsion and used its tail as a rudder; it did not yet have flukes or tail propulsion similar to modern whales.

The year after *Ambulocetus* and *Rodhocetus* were found, scientists in Pakistan found *Dalanistes*. It is even more whale-like than *Ambulocetus*, with smaller but functional front and hind legs, but a much larger snout, more like an archaeocete whale. This made it a perfect intermediate between *Ambulocetus* and *Rodhocetus*. And *Rodhocetus*, in turn, is a beautiful transitional fossil linking more primitive whales with archaeocetes.

In the years since the 1990s, more and more transitional whale fossils have been found. *Takracetus* and *Gaviocetus* have progressively more reduced hands and feet, developing into whale-like flippers (see figure 10.5). Their bodies are also more dolphin-like, with further development of tail propulsion (as in living whales), meaning they probably had horizontal tail flukes as well. So many transitional whale fossils have been found now that it's impossible to decide where terrestrial animals end and true whales begin. Whales were a complete mystery with no good transitional fossils in 1980, but today the origin of whales from land animals is one of the best documented evolutionary transitions in the fossil record.

Finally, the huge middle Eocene archaeocete whales first found in Alabama were common in the marine beds of Egypt, just west of the pyramids. In 1990, Gingerich and colleagues were collecting these fossils when they found complete articulated skeletons, not just the jumble of bones found in Alabama. When they looked at the back of the skeleton, they found it still had functional hind limbs that stuck out from its body. But these hind legs were only as large as a man's arm, tiny for a 50-foot whale, and were not much use for walking for a completely aquatic animal of this size. They were vestigial remnants of limbs that had nearly, but not completely, vanished. If you look at a mounted skeleton of a living whale (see figure 10.3) with all the bones put in the right place, you'll find tiny hip bones and thighbones in the region around the backbone where the hips used to be. They are floating in the tissues of the whale, no longer performing any function, and are mute witnesses to the days when all whales had hips and hind legs and could walk.

The archaeocete whales that dominated the middle and late Eocene oceans then evolved into the two branches of whales that still survive today. One group is the toothed whales, or odontocetes, including dolphins, porpoises, and sperm whales. They have many specialized adaptations not found in the ancient archaeocetes, but at least they still have teeth. The more specialized living whales are the baleen whales, or mysticetes, the humpback whale, the gray whale, the fin whale, and many others, including the largest animal to ever live—the mighty blue whale. These whales have entirely lost their teeth, and their upper jaws are lined with a screen-like fibrous filtering device called *baleen*, which is made of the same keratin that is found in your hair and fingernails. Even though they are the largest animals on the planet, they feed on some of the tiniest, especially planktonic crustaceans known as krill. They feed by opening their mouths wide and gulping a huge volume of krill-laden seawater into their mouth cavity. Then they force the seawater back out through the baleen filtering device, using their huge tongues and collapsing their big throat muscles to squeeze out the water. The krill are trapped in the baleen, ready to be swallowed.

These whales are truly amazing creatures, but it is difficult to imagine them evolving from toothed whales. But those fossils have been found too. The mouth of *Llanocetus*, from the late Eocene of Antarctica, has both ancestral triangular archaeocete teeth and baleen at the same time. By the late Oligocene, there were even more advanced toothed mysticetes such as *Janjucetus* and *Mammalodon* from Australia. In the early Oligocene, an archaic family of toothed mysticetes, the aetiocetids, appeared in Australia, but they became common whales in the North Pacific region (Japan and North America) during the late Oligocene to Pleistocene. Meanwhile, the first fossils of toothless mysticetes (*Eomysticetus*) appeared in the late Oligocene of South Carolina. That's why even today baleen whales develop tiny teeth during their early embryology, but lose them when they become fully adult. This is another example of a vestigial organ in whales that goes along with the tiny hind limbs and hips.

The last piece of this puzzle came from molecular biology. Even in the earliest days of analysis by the old methods of immunological distance, or DNA hybridization, whales were consistently found to be related to the even-toed hoofed mammals, or artiodactyls. When protein sequences and then actual DNA sequences were analyzed in the 1990s, it became even more clear that whales were a group *within* artiodactyls, not their own separate order independent of other mammals, as they had been classified for so long. Again and again the molecular data pointed to one group of artiodactyls: the hippopotamuses (see figure 10.5). This was consistent with fossils showing that whales were closely related to an extinct group of pig-like terrestrial mammals called the anthracotheres, which were also ancestral to hippos.

Paleontology gave us the final piece of evidence proving that whales were artiodactyls. In 2001, two independent groups of paleontologists (Thewissen's group and Gingerich's group) found and reported fossils of early whales in Pakistan that had well-preserved ankles (figure 10.6). In both cases, the ankle bones of these earliest whales had the diagnostic "double-pulley" configuration of the astragalus bone (the hinge bone in the mammalian ankle joint), which was originally known only in the artiodactyls. Unlike any other group of mammals, all artiodactyls have this

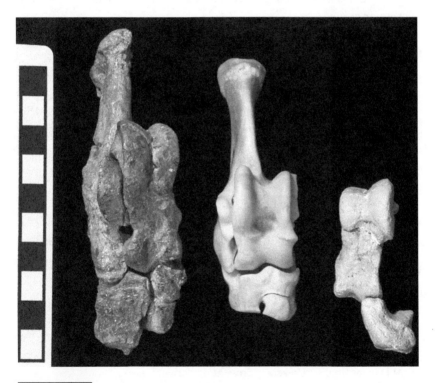

Figure 10.6 ▲

Ankle bones of middle Eocene whale *Rodhocetus balochistanensis* (*left*) and *Artiocetus clavis* (*right*) from Pakistan, compared to those of the pronghorn *Antilocapra americana* (*center*). Note that whales had the double-pulley astragalus characteristic of mammals of the order Artiodactyla. (Courtesy of Philip Gingerich)

double-pulley hinged ankle bone, and indeed, most artiodactyls can be identified as members of that order by this unique bone alone. Looking at both of these fossils, it was clear that whales had the unique anatomy of the artiodactyl ankle as well.

And so we reach this conclusion. From mammals once mistaken for fish, we now have lots and lots of transitional fossils that show whales evolving from the anthracotheres to *Indohyus* to *Pakicetus* to *Ambulocetus* and many more transitional forms. We have fossils and molecular evidence that show whales are descended from the artiodactyls, the group that included the anthracotheres. And we have Darwin's evidence pointing to vestigial organs such as tiny relict hips and thighbones that originally suggested whales came from land animals with fully functional hind legs for walking, not swimming. Vestigial organs are indeed the mute witnesses of the past, as Darwin argued.

NOTE

1. Edward J. Behrman, George A. Marzluf, and Ronald Bentley, "Evidence from Biochemical Pathways in Favor of Unfinished Evolution Rather Than Intelligent Design," *Journal of Chemical Education* 81, no. 7 (2004): 1051–1052.

FOR FURTHER READING

Berta, Annalisa. *The Rise of Marine Mammals: 50 Million Years of Evolution*. Baltimore, Md.: Johns Hopkins University Press, 2017.

Berta, Annalisa, James L. Sumich, and Kit M. Kovacs. *Marine Mammals: Evolutionary Biology*. 3rd ed. Amsterdam: Academic Press, 2015.

Gould, Stephen Jay. "Hooking Leviathan by Its Past." *Dinosaurs in a Haystack: Reflections in Natural History*. New York: Norton, 1997.

Prothero, Donald R. *The Princeton Field Guide to Prehistoric Mammals*. Princeton, NJ: Princeton University Press, 2016.

Prothero, Donald R., and Scott E. Foss, eds. *The Evolution of Artiodactyls*. Baltimore, Md.: Johns Hopkins University Press, 2007.

Prothero, Donald R., and Robert M. Schoch. *Horns, Tusks, and Flippers: The Evolution of Hoofed Mammals*. Baltimore, Md.: Johns Hopkins University Press, 2002.

Pyenson, Nick D. *Spying on Whales: The Past, Present, and Future of the World's Most Awesome Creatures*. New York: Viking, 2018.

Savage, R. J. G., and M. R. Long. *Mammal Evolution: An Illustrated Guide*. New York: Facts-on-File, 1986.

Thewissen, J. G. M., ed. *The Emergence of Whales: Evolutionary Patterns in the Origin of Cetacea*. New York: Plenum Press, 1998.

——. *The Walking Whales: From Land to Water in Eight Million Years*. Berkeley: University of California Press, 2014.

Zimmer, Carl. *At the Water's Edge: Macroevolution and the Transformation of Life*. New York: Free Press, 1998.

INVASION OF THE LAND

Typical summers of my adult life are spent in snow and sleet, cracking rocks on cliffs well north of the Arctic Circle. Most the time I freeze, get blisters, and find absolutely nothing. If I have any luck, I find ancient fish bones. That may not sound like buried treasure to most people, but to me it is more valuable than gold. Ancient fish bones can be a path to knowledge about who we are and how we got that way. We learn about our bodies in seemingly bizarre places, from the fossils of worms and fish recovered from rocks around the world to the DNA in virtually every animal alive on earth today.

—NEIL SHUBIN, *YOUR INNER FISH: A JOURNEY INTO THE 3.5-BILLION-YEAR HISTORY OF THE HUMAN BODY* (2008)

Two scientists were crouched in the bitter Arctic chill, looking hard at the ground for signs of fossil bone. Even though it was high noon, the temperature in this remote part of the Canadian Arctic was barely above freezing during the warmest part of the day, and the winds never stopped blowing across the barren tundra landscape. Not only were they wearing bright orange polar parkas and ski masks and goggles to protect against snow blindness, windburn, and frostbite, but they also carried rifles in case they stumbled across a hungry polar bear. It was just after the first day of summer, and the "midnight sun" circled low on the horizon all day long and never set. To sleep, they had to force themselves to go to bed when the clock said it was night, even though it was still daylight outside.

The two scientists, my friends Neil Shubin of the University of Chicago and Ted Daeschler of the Philadelphia Academy of Natural Sciences, were there because these reddish sandstone outcrops promised the possibility of an important discovery. They were looking for a fossil intermediate in age between the earliest amphibians that had been collected but were not as advanced as the next youngest amphibian fossils known. They knew that previous discoveries of fossils during that time period had been found in rocks from the Upper Devonian. So they looked on geologic maps for rocks covering a certain time span (385–365 million years ago) within the Late Devonian Period. Good fossils of very primitive amphibian-like fish had already been found from 385 million years ago, and fossils of more advanced amphibians had been found dating from 365 million years ago. After scouring the geologic maps of the world, they found only three areas of the right age and the right sedimentary environment (shallow marine sandstones and shales, formed in rivers or deltas). Two of these areas had already been explored, one in Pennsylvania, where some very advanced fish had been found, and one in Spitsbergen and Greenland in the Arctic, which produced the well-known amphibians *Acanthostega* and *Ichthyostega*. A third place, in the Canadian Arctic, had never been studied, and it dated between 385 and 365 million years ago.

They raised grant money for a quick visit to the area in 1999, and they found some promising bone scraps. Then they had to raise millions of dollars to mount a full-scale Arctic expedition, which they did for several summers in a row (figure 11.1A; figure 11.1B). As Shubin described it in his book *Your Inner Fish*:

> I first saw one of our inner fish on a snowy July afternoon while studying 375 million year old rocks on Ellesmere Island, at a latitude about 80 degrees north. My colleagues and I had traveled up to this desolate part of the world to try to discover one of the key stages in the shift from fish to land-living animals. Sticking out of the rocks was the snout of a fish. And not just any fish: a fish with a flat head. Once we saw the flat head, we knew we were on to something. If more of this skeleton were found inside the cliff, it would reveal early stages in the history of our skull, our neck, and our limbs. (8–9)

Only after the third year of very hard, expensive, dangerous work dealing with harsh weather and marauding polar bears did they find fossils of *Tiktaalik*, a critical transitional fossil between fish and amphibians (figure 11.2A; figure 11.2B). The name *Tiktaalik* is the Inuit word for a local freshwater fish hunted by the people of the region. As codiscoverer Ted Daeschler said, "We found something that really split the difference right

Figure 11.1 ▲

Neil Shubin, Ted Daeschler, and crew collecting in the Devonian rocks of the Canadian Arctic. (*A*) They required polar-grade tents able to stay up under howling winds and blizzards, as well as lots of gear normally not required outside the polar regions. (*B*) The *Tiktaalik* quarry after a season of excavation. (Courtesy of Neil Shubin)

Figure 11.2 ▲

(*A*) The skeleton and (*B*) a life-sized reconstruction of *Tiktaalik*. (Courtesy of Neil Shubin)

down the middle." Fish-amphibian expert Jenny Clack wrote of their discovery, "It's one of those things you can point to and say, 'I told you this would exist,' and there it is."

The lobed fins of *Tiktaalik* had all the elements ancestral to the amphibian limb, but it still had fin rays rather than toes (figure 11.3). It had fish-like scales, a combination of both gills (shown by the gill arch bones) and lungs (shown by the spiracles in its head), and a fish-like lower jaw and palate. Unlike any fish, it had amphibian features too: a shortened flattened skull

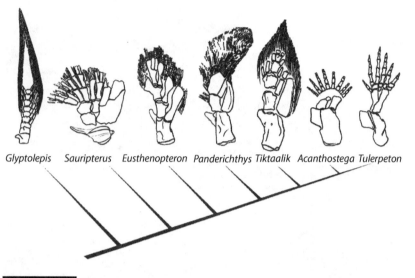

Glyptolepis   Sauripterus   Eusthenopteron   Panderichthys   Tiktaalik   Acanthostega   Tulerpeton

Figure 11.3 ▲

The transformation of the pectoral fin of lobefins into the hand and forelimb of primitive tetrapods. Each bone of the lobefin is homologous with one of the limb bones of the tetrapod. The main differences are modifications in shape and robustness and replacement of the fin rays with fingers. (From Neil H. Shubin, Edward B. Daeschler, and Farish A. Jenkins Jr., "The Pectoral Fin of *Tiktaalik roseae* and the Origins of the Tetrapod Limb," *Nature* 440 [2006]: figure 4, used by permission of Nature Publishing Group.)

with a mobile neck, notches for the eardrums on the back edge of the skull, and robust ribs and limbs and shoulder and hip bones. Yet, like *Acanthostega*, its fins were not strong enough or flexible enough to allow it to crawl on land; instead, the fins were probably used to paddle in shallow water and push up so it could see above the surface. Like the other transitional fish-amphibian fossils (and many modern amphibians, especially newts and salamanders), it probably spent most of its time in the water, using its limbs to push along and paddle. It could hunt on the margins of the streams in which it lived, but it was not capable of dragging itself across the land very far, or walking with its belly off the ground.

After this spectacular find, Shubin and Daeschler and their crew returned several more times, finding many more specimens of *Tiktaalik* along with a number of other fish and animals that lived in this ancient river delta around 375 million years ago. The discovery of *Tiktaalik* shows the predictive power of geology and evolution. Shubin and his colleagues already knew what time interval to look for (375 million years ago) based on the age of the most advanced fish-like transitional fossils and the most primitive amphibians that had already been discovered. They consulted geological maps to find rocks of the right age and type, eliminating those areas that had already been explored, and located the exposures of rock sequences in the Canadian Arctic in which the crucial fossils were discovered, providing the "missing link" in the evolutionary chain.

Neil and Ted's excellent adventure was just the latest find in a long history of searching for fossils that connected fish with land vertebrates. Even early naturalists had noticed that the lungfish bore similarities to amphibians, but they were still fish with fins. Their "lobed fins" were not made of a fan of long, rod-like fin rays similar to those of most living fish. Instead, the lobed fins were made of a series of bones that closely matched the bones in the vertebrate arm and leg. In fact, when the first South American lungfish was discovered and described in 1837, it bore useless ribbon-like fins that trailed behind it as it swam like an eel. Some naturalists thought it was a degenerate amphibian because it had lungs. But when the African lungfish was discovered and described, its lobed fins had the same bones as that of land amphibians. Eventually, lungfish fossils were found with this kind of robust bony fin, not the bizarrely modified fins of the South American lungfish.

Bit by bit, more and more fossils were discovered that the bridged the once large gap between fish and amphibians. In 1882, Joseph P. Whiteaves published a two-paragraph description of *Eusthenopteron foordi* that mentioned almost none of its amphibian-like features and had no illustrations. It was a big (up to 1.8 meters [6.5 feet] long) lobe-finned fish that was more like an amphibian than any living lungfish or coelacanth. It is now known from many beautiful specimens from a famous fossil locality called Miguasha, on Scaumenac Bay, Quebec. It had all the right bones from which to construct the amphibian arm and leg and all the right bones in the skull to be ancestral to amphibians.

In the early twentieth century, more and more Late Devonian lobed-finned fish fossils were discovered, and lots of Early Carboniferous amphibian fossils followed, but there were no fossils of the transition of amphibians to fish yet. Then, in 1920, unrelated political forces accidentally led to the discovery of key fossils. During the 1920s, Denmark and Norway were disputing the ownership of East Greenland, which their Viking ancestors had visited over a thousand years earlier. Neither country, however, had explored the region much, so it was imperative to do some research on the region to establish their claim. Consequently, the Danes funded a three-year expedition in the summers of 1931 to 1933, funded by the Carlsberg brewery and the Danish government. Led by the famous Danish explorer and geologist Lauge Koch, it was staffed by prominent Swedish and Danish botanists, zoologists, geologists, geographers, and archeologists—and Swedish paleontologist Gunnar Säve-Söderbergh, only 21 years old when he joined the expedition. He found fossils of many remarkable creatures such as *Ichthyostega* and *Acanthostega*, many lungfish, and *Osteolepis*, a lobed-finned fish much like *Eusthenopteron*. During the 1920s and 1930s, Säve-Söderbergh wrote short descriptions and named his fossils, but he died in 1948 at the age of 38 of tuberculosis without having completed a detailed analysis. Eventually his specimens were studied by Swedish paleontologist Erik Jarvik, who had been with Säve-Söderbergh on some of the later expeditions. Known to be a slow, meticulous worker, Jarvik did detailed analyses of the fossils, but took *50 years* before he finally published his description of all the fossils of *Ichthyostega* in 1996—when he was 89 years old!

Until the details were finally published by Jarvik, Säve-Söderbergh's sole illustration of *Ichthyostega* was the only published transitional fish-amphibian fossil available for comparison for many years. It was a remarkable transitional fossil nonetheless, with an interesting mixture of fish-like and amphibian-like features (figure 11.4). Like a fish, it still had a large tail fin for underwater propulsion, large gill slits on the side of the head, and the network of closed canals on the face, called lateral lines, that most fish use to sense changes in the water currents around them. Yet, like an amphibian, it clearly had well-developed arms and legs with fingers and toes that would propel it across a hard surface. (Later research has shown that the forelimbs were not strong enough to do much walking, so it moved in short hops, dragging its flipper-like hind limbs). Like modern newts and salamanders, its limbs were mostly used for pushing through obstacles in the water, not for lifting their body above the ground in fast walking. The ribs of *Ichthyostega* had robust flanges on them, supporting its chest cavity for breathing out of water—but preventing it from the rib-propelled breathing used by many amphibians. It also had a long flat snout with eyes that looked upward and a short braincase, in contrast to the deep, cylindrical skull of *Eusthenopteron* and many other lobed-finned fish that had a short snout and large braincase and eyes that faced sideways.

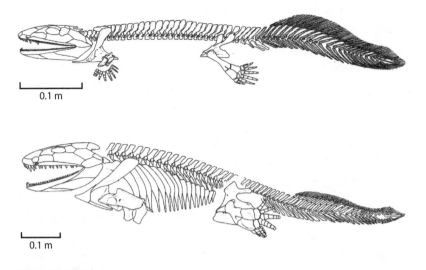

0.1 m

0.1 m

Figure 11.4 ▲

Sketch of the skeletons of *Acanthostega* (*top*) and *Ichthyostega* (*bottom*), showing the mixture of fish-like features (tail fins, lateral line systems, and gill slits) and tetrapod features (robust limbs and shoulder and hip bones, reduced back of skull, and expansion of the snout). (Drawing by Michael Coates)

In the 1980s and 1990s, research on transitional fossils between fish and amphibians shifted from Sweden to Cambridge University, with Jenny Clack, Per Ahlberg, Michael Coates, and their students. They realized that there must be much more fossil material in Greenland than the limited material collected by the Danes in the 1920s. As Clack wrote in 2002:

> In 1985, I began to think about the possibility of an expedition to East Greenland, at the instigation of my husband Rob. Along the trail, I met Peter Friend of the Earth Sciences Department across the road in Cambridge, who had been leader of several expeditions to the part of Greenland in which I was interested. It turned out that he'd had a student, John Nicholson, who'd collected a few fossils as part of his thesis work on the sediments of the Upper Devonian of East Greenland between 1968 and 1970. Peter retrieved these specimens from a basement drawer and also showed me John's notebook from his 1970 expedition. John's note that on Stensiö Bjerg, at 800 metres, *Ichthyostega* skull bones were common was startling, and portentous. The fossils that he'd collected fitted together to make a single small block of three

partial skulls and shoulder girdle bits—not of *Ichthyostega*, but of its at that time lesser known contemporary, *Acanthostega*. Peter suggested I get in touch with Svend Bendix-Almgreen, Curator of Vertebrate Palaeontology in the Geological Museum in Copenhagen. The Danes still administered expeditions by geologists to the National Park of East Greenland, where the Devonian sites are located, so he would be the person to start with in my attempts to mount an expedition there. Peter also suggested I contact Niels Henricksen of the Greenland Geological Survey (GGU). By sheer coincidence, and great good fortune, the GGU had a project in hand in the very place where I needed to go, and their last season there was the summer of 1987. With funds from the University Museum of Zoology and the Hans Gadow Fund in Cambridge and the Carlsberg Foundation in Copenhagen, I, my husband Rob, my student at the time, Per Ahlberg, and Svend Bendix-Almgreen and his student Birger Jorgenson arranged a six-week field trip in the care of the GGU for July and August of 1987. Using John Nicholson's field notes, we eventually pinned down the locality from which the *Acanthostega* specimens had come, and then the exact in-situ horizon that had been yielding them. It was in effect, a tiny, but very rich, *Acanthostega* "quarry."

Clack's group found many of the same fossils as Säve-Söderbergh, but they also found much better material of a poorly known fossil named *Acanthostega* by Jarvik in 1952 (see figure 11.4). *Acanthostega* was much more fish-like than *Ichthyostega*, with more fish-like limbs indicating that it could never have crawled on land, and lacking wrists, elbows, or knees. Although it still had arms and legs rather than fins, they were more for pulling itself through the water and could not propel it across land much. The biggest shock was the more complete hands and feet that were poorly known in *Ichthyostega*. It had as many as seven or eight fingers on the hands, not the five fingers that most vertebrates bear, but only four toes on its feet. It was also more fish-like in having a very large tail fin and short ribs that would prevent it from breathing on land. Yet it was still like many amphibians in having ears that could hear in water and on land, strong bones in the hips and shoulders, and a neck joint that allowed it to swing its head to the side quickly and snap at prey, which was unlike any fish.

More and more discoveries continued to be made, and the family tree of the fish to amphibian transition is remarkably complete today (figure 11.5). Moving up from primitive lobefins such as lungfish and coelacanths, we

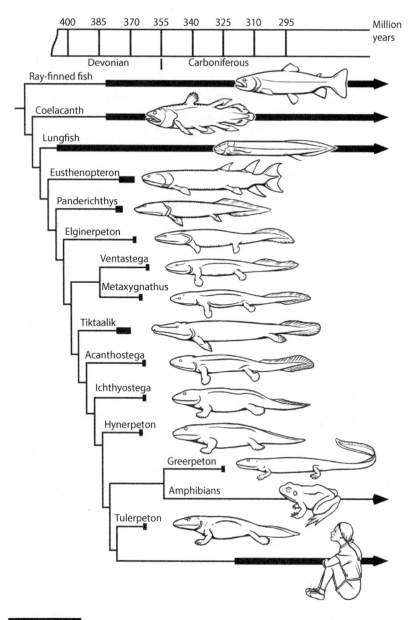

Figure 11.5 ▲

Phylogeny of the transitional series from "rhipidistians" through primitive tetrapods. (Drawing by Carl Buell)

have the very fish-like lobed-finned *Eusthenopteron*, and slightly more advanced fossils called *Panderichthys, Elginerpeton, Ventastega*, and *Metaxygnathus*. *Tiktaalik* is slightly more amphibian-like but still does not have hands or feet like *Acanthostega*, and finally the most amphibian-like fossil, *Ichthyostega*. From there, we have fossils that everyone recognizes as amphibians. Once could not ask for a more complete transitional sequence between two major groups of animals, although there are many such transitional sequences known now.

The fossil record now shows how and when and where amphibians first crawled out onto land. But why did they do it? What drove fish to make the difficult transition onto land, where they needed new sense organs, new ways of respiration (lungs instead of gills), and stronger limbs and rib cages and more robust shoulder and hip bones fused to the spine to support their weight out of water? For decades, scientists discussed the topic as if it were some implausible miracle. But, in fact, many animals that have partially made that transition can be seen all around us. The ray-finned fish (99 percent of living fish, including most of the fish you eat or have in your fish tank) have done it independently many times in many different groups. For example, "walking catfish" (figure 11.6A) can wriggle across the land from one pool to another when their home begins to dry up, the water becomes foul, or to find a new pool with new food resources when the old pool is too crowded. Climbing perch wriggle and crawl across dry land to find better pools; they can even crawl up trees, hence their name. Mudskippers (figure 11.6B) live permanently right on the boundary of land and water. They graze on algae and prop themselves up on mangrove swamps and mud flats by their ray fins, and they use their stalked eyes to see out of the water when they are submerged. They can flee to water when predators threaten from land, and to land when predators appear in the water. Many tidepool fish, such as gobies and sculpins, spend much of the low tide crawling along the rocks with their hand-like fins, preying on animals trapped by the low tide. Spotted moray eels wriggle out of the water during low tides to prey on crabs that are looking for smaller food to eat. A number of other fish have modified the fin-rays of their front fins into clumsy "fingers" that enable them to crawl across surfaces (figure 11.6C).

These are all ray-finned fish, not closely related to lungfish or coelacanths or the other lobe-finned fishes that gave rise to amphibians. All of these examples of fish with semiterrestrial lifestyles evolved independently

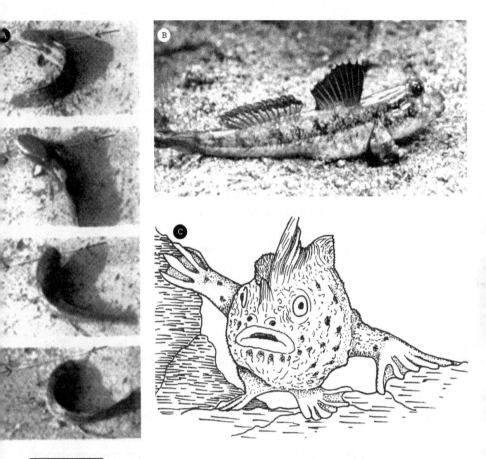

Figure 11.6 ▲

A number of ray-finned fishes have evolved the ability to live on land and crawl around, or they have modified their rayed fins into walking appendages for use in creeping along the seafloor. (*A*) The "walking catfish" wriggles along the ground between ponds when its home pond dries up or becomes too crowded. (*B*) The mudskipper spends most of its life out of water sitting on the mudflats or mangrove roots. (*C*) The frogfish has modified its ray fins into "fingers" that enable it to creep along the bottom. ([*A*] After Alfred Sherwood Romer, *The Vertebrate Story* [Chicago: University of Chicago Press, 1959]; [*B*] courtesy of Wikimedia Commons; [*C*] Jennifer A. Clack, *Gaining Ground: The Origin and Early Evolution of Tetrapods* [Bloomington: Indiana University Press, 2002], figure 4.15, used with permission.)

in multiple groups, all in different ways. Clearly there are strong pressures for fish to exploit the land (at least for short periods of time) to find new food or to escape predators or crowding in the water. It is also clear that it is no big deal for fish to do this because entirely unrelated groups of fish have evolved to do this many different times and to different degrees. Instead of the difficulties scientists had imagined just a few decades ago, it now seems a trivial task because it was done so often by so many different unrelated groups of fish. As Shubin wrote in *Your Inner Fish*:

> What possessed fish to get out of the water or live in the margins? Think of this: virtually every fish swimming in these 375-million-year-old streams was a predator of some kind. Some were up to sixteen feet long, almost twice the size of the largest *Tiktaalik*. The most common fish species we find alongside *Tiktaalik* is seven feet long and has a head as wide as a basketball. The teeth are barbs the size of railroad spikes. Would you want to swim in these ancient streams? (41)

Finally, a recent study by a group of scientists led by Emily Standen showed just how easy it is for fish to become modified for at least some kind of land life. The bichir, found in Africa (genus *Polypterus*) and distantly related to sturgeons and paddlefish, is a very primitive ray-finned fish that has some similarities to the earliest fossil lobefins. These bichirs were raised on land rather than in water (they are already good air breathers). In just a few generations of breeding, their fins became more robust and better for land crawling than those of their ancestors. Clearly, the genes for modifying fins into something else are easy to trigger, and this mechanism was employed by many of the land-living ray-finned fish we have discussed.

If anyone says that they cannot imagine fish crawling out onto land and becoming amphibians, you can tell them that they only need look at the fossil record and at the behavior of many living ray-finned fish to find the answer.

FOR FURTHER READING

Clack, Jennifer A. *Gaining Ground: The Origin and Early Evolution of Tetrapods*. Bloomington: Indiana University Press, 2002.

Daeschler, Edward B., Neil H. Shubin, and Farish A. Jenkins Jr. "A Devonian Tetrapod-Like Fish and the Evolution of the Tetrapod Body Plan." *Nature* 440 (2006): 757–763.

Long, John A. *The Rise of Fishes*. Baltimore, Md.: Johns Hopkins University Press, 1995.

Maisey, John G. *Discovering Fossil Fishes*. New York: Henry Holt, 1996.

Shubin, Neil. *Your Inner Fish: A Journey Into the 3.5-Billion-Year History of the Human Body*. New York: Pantheon Books, 2008.

Shubin, Neil H., Edward B. Daeschler, and Farish A. Jenkins Jr. "The Pectoral Fin of *Tiktaalik roseae* and the Origins of the Tetrapod Limb." *Nature* 440 (2006): 764–771.

Zimmer, Carl. *At the Water's Edge: Macroevolution and the Transformation of Life*. New York: Free Press, 1998.

MISSING LINKS FOUND

Natural selection is daily and hourly scrutinizing, throughout the world, every variation, even the slightest; rejecting that which is bad, preserving and adding up all that is good; silently and insensibly working. . . . We see nothing of these slow changes in progress until the hand of time has marked the long lapse of ages. . . . Why then is not every geological formation and every stratum full of such intermediate links? Geology assuredly does not reveal any such finely graduated organic chain; and this, perhaps, is the gravest objection which can be urged against my theory.

—CHARLES DARWIN, *ON THE ORIGIN OF SPECIES* (1859)

One of the chief complaints critics of evolution often level at scientists is, "Where are the missing links? Show me a fossil that demonstrates the transition of one major group to another." These critics often concede that microevolution occurs, from small changes in the fruit fly to pesticide resistance in insects to the rapid changes in viruses and bacteria. These changes happen in experiments in real time, so they cannot be denied. But that is not good enough for critics of evolution. They balk at the idea that major distinct groups are closely related (such as birds and reptiles) and claim that missing links do not exist that show how this transition could have occurred. Although they admit that changes on the scale of fruit flies and microbes are happening, this is dismissed as microevolution. And they continue to assert that the big changes between major groups—macroevolution—are impossible.

First, let's clear up some misconceptions. There are no such things as "missing links" because the entire concept is erroneous and 160 years out of date. The idea of missing links dates back to Aristotle's thinking more than 2,000 years ago, when life was envisioned as a "great chain of being," or "ladder of creation" (*scala naturae* in Latin). Each type of organism was placed on a separate rung of the ladder, with plants and sponges and corals at the bottom, insects and fish near the middle, mammals near the top, and, of course, humans at the very top (figure 12.1). In some religious versions of this thinking, the cherubim and seraphim and lower angels were just above humans, followed by archangels, with God at the top. That is the origin of

Figure 12.1 ▲

Evolution is *not* about life climbing the "ladder of nature" or finding the "missing links" in the "great chain of being" from "lower" to "higher" organisms. Instead, evolution is a "bush" with many lineages branching from one another, and ancestors living alongside their descendants. (Drawing by Carl Buell)

the outdated terminology of "lower organisms" and "higher organisms." Likewise, some naturalists thought of sponges and plants as one end of the chain of being, with humans at the other end, so whenever no organisms were known that seemed to complete the chain, we had a missing link.

But that entire concept of life became obsolete in 1859 when Darwin showed that life has a branching, bushy history—not a single linear trend through time. Most organisms are parts of a "family tree" with many different branches, some of which overlap each other in time and space. This is especially true of the archaic linear thinking regarding the evolution of the horse (chapter 14) and the evolution of humans (chapter 24). In both cases, the familiar diagram showing the "march of progress" from primitive horse to modern *Equus* or from a chimp to modern *Homo sapiens* gives the appearance of a single lineage with no branching. This "ladder of life" concept was possible for horses or humans because only one genus of horse is alive today and only one species of human is alive today, so you can draw a straight line between these isolated endpoints through any number of intermediates. This linear model of horse and human evolution became iconic images and were repeated endlessly in textbooks and in the media (figure 12.2A; figure 12.2B). In contrast, many kinds of living animals can be found in most groups alive today, making it impossible to shoehorn all their intermediate forms into a simple linear model leading to a single living descendant. When someone asks me to show them a missing link, I reply that the concept is outdated, misleading, and meaningless—but I am happy to show them lots of transitional fossils that connect the major living groups of animals.

So what about transitional fossils? Are there no fossils that demonstrate the ancestry of one group from another, as many science deniers claim? On the contrary, the fossil record is loaded with such examples, and I described many of them in my 2017 book, *Evolution*. Some of those examples are discussed at length in other chapters in this book, such as whales (chapter 10), amphibians from fish (chapter 11), dinosaurs from birds (chapter 13), horses (chapter 14), giraffes (chapter 15), and elephants (chapter 16). Many additional examples could be given, but I'll limit this discussion to just a few striking examples to make the point that there is no shortage of fossils that show how one major group made the transition from another.

First, let's look at the frog. Everyone knows these from the media, and probably most people have seen one in the wild and some may have dissected one in a biology class at one time or another. A frog looks highly

FIG. 204.
(Cf. p. 181.)
Gibbon.

FIG. 205.
(Plate XIV. Fig. 3.)
Orang-outang.

FIG. 206.
(Plate XIV. Fig. 1.)
Chimpanzee.

FIG. 207.
(Plate XIV. Fig. 2.)
Gorilla.

FIG. 208.
(Plate XIV. Fig. 4.)
Man.

Figure 12.2 ▲

(A) One of many versions of the iconic "march of progress" from apes to man; this one is from Giuseppe Donatiello's *Astronomy Evolution*. (B) The famous 1863 diagram from Thomas Henry Huxley, showing the skeletal similarities of humans and apes, which was the foundation for the erroneous "apes to man" icon. ([A] Courtesy of Wikimedia Commons; [B] Thomas Henry Huxley, *Evidence as to Man's Place in Nature* [London: Williams & Norgate, 1863])

specialized with its huge broad mouth and long protrusible sticky tongue, large eyes, short trunk with tiny ribs, very elongated hip bones, extremely long hind legs, and no tail. Frogs cannot use their ribs for breathing; instead, an inflatable pouch in the frog's throat pumps air in and out (it is also used for making a variety of sounds). Frogs range tremendously in size, from the tiny New Guinean frog (only 7.7 millimeters [0.3 inches] long) to the Goliath frog, which is over 300 millimeters (12 inches) long and weighs 3 kilograms (7 pounds). It is so big that it eats birds and small mammals as well as insects.

It's hard to imagine an intermediate fossil between it and more typical amphibians such as salamanders. Yet we can see this transition in two ways. First, we know that frogs make this transition during their own ontogeny through their development from tadpoles to adults. When a tadpole still has its tail and first develops limbs, its legs are relatively equal in length and do not yet show the extreme proportions seen in adult frogs. Before the tail is resorbed, a tadpole goes through a salamander-like stage.

Even more striking is the fossil record of this transition. For decades, pale-ontologists had an excellent fossil record of amphibians that resembled sal-amanders, although some were huge. The biggest ones were from the Lower Permian red beds of north Texas and southern Oklahoma, especially around the tiny town of Seymour, Texas. One of these huge amphibians was *Eryops*, a crocodile-shaped fossil known from numerous complete skeletons. It had a large sprawling body over 2 meters (6.5 feet) long, with a robust tail and limbs, and a skull well over 60 centimeters (2 feet) long in some individuals! This was one of the largest terrestrial animals of the Early Permian, capable of hunting prey both in the water and on the land. The slightly more primi-tive *Edops* from Early Permian red beds of Texas had an even longer skull, and it would have been even larger than *Eryops* if its complete skeleton were known. These fossils are abundant in the Early Permian red beds, along with many other weird groups of amphibians, plus early reptiles and the earliest members of the mammalian lineage (such as the finbacked *Dimetrodon*, which is found in all the children's dinosaur kits and books but is related to us, not dinosaurs). Paleontologists who work on early reptiles and amphibi-ans all spent time collecting in the Texas Permian red beds, finding mostly more and more fossils of *Dimetrodon*, *Eryops*, and other familiar animals.

One of these collectors was my friend, the late Nick Hotton, who spent most of his career at the Smithsonian. In 1994, one of his crews was collect-ing in a locality nicknamed "Don's Dump Fish Quarry." There were plenty of fossil fish and amphibians, but they had time only to quickly collect the fossils and did not do a detailed study in the field. Nick Hotton recognized the importance of one particular fossil (found by Peter Krohler, a curatorial assistant at the Smithsonian). He put it in his pocket with a slip of paper that said "Froggie" on it. When he died in 1999, he still had not gotten a chance to study it or publish it. So the specimen sat unstudied for 14 years until a younger generation of paleontologists, led by Jason Anderson of the University of Calgary, Robert Reisz of the University of Toronto, and others retrieved it from the collections, finished cleaning the matrix off of it so it was completely exposed, and finally published it. They named it *Gerobatrachus hottoni* (Hotton's ancient frog), but the press called it the "Frogamander" as news of its discovery spread (figure 12.3A; figure 12.3B).

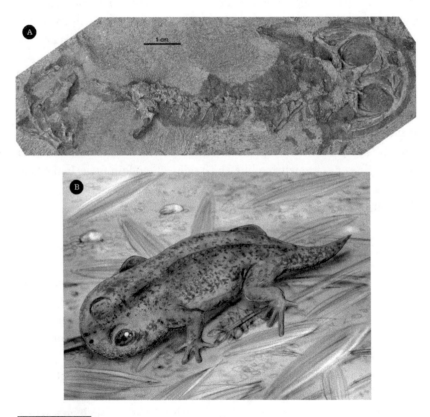

Figure 12.3 ▲

(A) The only specimen of *Gerobatrachus hottoni*. (B) A reconstruction of it in life. ([A] Courtesy of Diane Scott and Jason Anderson; [B] courtesy of Nobumichi Tamura)

At first glance, the specimen seems unimpressive; it is nearly complete but only about 11 centimeters (4.3 inches) long. It is lying on its back with only a part of the hip bones, shoulder bones, and tail missing. The most striking thing about the fossil is that it combines the long-tailed body of a salamander with the short, broad, rounded snout of a frog, showing how frogs might have begun to evolve from salamander-like forms. It has a few other froggy features: a large eardrum and teeth that sit on tiny pedestals with a distinct base, an anatomical condition found only in the living amphibians and their close fossil relatives. Otherwise it has the primitive salamander-like body, so it is a perfect transition between the two groups.

After the Early Permian *Gerobatrachus*, the next good frog fossil is *Triadobatrachus* from the Early Triassic (240 million years old) of Madagascar. It has the typical froggy broad snout and long webbed feet, but unlike any living frog, it still has a long trunk region, with 14 vertebrae in its spine, not the 4 to 9 found in living frogs. It even retains a short tail that is not lost when its tadpoles grew to adulthood. It had longer hind legs than any salamander, but not the huge muscular legs found in living frogs, so it could swim but it could not jump. By the Early Jurassic (about 200 million years ago), we have fossils of the first true frog, *Vieraella*, from Argentina. A tiny creature only 3 centimeters (2 inches) long, its skull was completely frog-like, the hind limbs were capable of jumping, but it did not have the short trunk region or extremely modified hip region of modern frogs. In the Cretaceous, frogs looked almost completely modern in their anatomy and had diversified into many of the groups that are alive today, with dozens of families that include more than 5,700 living species.

As a second example, how about turtles? I've heard anti-evolutionists scoff at the idea that there even could be an intermediate fossil between a turtle and any other reptile. What good is a turtle with half a shell, they taunt? They look at pictures of all the fossil turtles that were known until recently and say "this is just another kind of turtle." Indeed, it seems hard to imagine a turtle without both of its shells. How would it even function? Would we even call it a turtle?

Turtles and their shells seem very stereotyped. Once turtles with complete shells appeared in the fossil record, it was such a successful body plan that it evolved and diversified, but always with a dome of shell on top (the carapace) and a belly shield (plastron). More than 1,200 species of turtles are alive today, and the roots of the modern groups can be traced back to the Jurassic. Most are easily separated into two categories: cryptodires and pleurodires. The cryptodires are the more familiar and most diverse group of turtles on the planet, making up all but about 80 species of living turtles. The name *cryptodire* means "hidden joint" and refers to the fact that the neck folds upon itself in a vertical S-bend inside the front of the shell when they pull their head into their shell. The much rarer and less diverse pleurodires (side joint), or side-necked turtles, fold their neck sideways, like closing a jackknife, and pull their head in sideways under the overhanging lip of the front of the shell. They are not only rare and endangered but are found today only on the Gondwana remnant continents of Australia, South America, and Africa.

At first glance, turtles seem to be a distinct kind of animal locked into a stereotypical body form that could never evolve from anything else. But transitional fossils show how turtles evolved from reptiles without shells. The first to be found was a strange fossil known as *Proganochelys*, from the Upper Triassic beds (210 million years old) of Germany, Greenland, and Thailand (figure 12.4). At first glance, it looks just like any other turtle, with a plastron and a carapace. However, a closer look shows that it is a lot more primitive and is not a member of any living group. For one thing, the carapace is very different, with many additional plates not seen in any living

Figure 12.4 ▲

Proganochelys, a Triassic turtle which had a fully formed plastron and carapace to form a shell, but could not withdraw its head, and still had some reptilian teeth on its palate. (Photograph by the author)

turtle, especially around the edge of the shell and protecting the legs. In addition, its tail is covered by a spiky bony sheath, with a spiky tail club. Even more primitive is the skull. It looked much more like one of the primitive Permian reptiles, not a turtle with its distinctive arrangement of jaw muscles and holes in the skull. Although it has a turtle's beak, the upper palate still had teeth, the last of the turtles to retain teeth. Most important of all, it could not retract its big head into its shell the way all living turtles do, so armor and spikes on top of its head protected it instead. To the unobservant person, it is "just a turtle," but it's completely unlike any living turtle because it has an unretractable neck and head and has teeth in its mouth.

In 2008, an astonishing collection of turtle fossils from the Late Triassic (210 million years ago) was announced from China (figure 12.5A; figure 12.5B). Known from dozens of complete specimens, it was given the formal scientific name *Odontochelys semitestacea* (toothed turtle with half a shell). It solves the riddle of how turtles got their shells. It has no shell or carapace on its back (just thick ribs), but it does have a plastron, or belly shield. It is literally a "turtle on the half-shell," a transitional form between modern turtles with both shells and its ancestors with no complete shell. Another completely un-turtle-like reptilian feature is a full set of teeth, similar to its ancestors but unlike the toothless beaks of modern turtles.

Odontochelys resolves another long-standing debate as well. For decades, some paleontologists argued that the turtle carapace comes from small plates of bone developed from its skin (*osteoderms*) that became fused together; others argued that the carapace was mostly made of expansions of its back ribs. *Odontochelys* shows that the latter position is correct. It had broadly expanded back ribs that are beginning to develop and connect into a shell, and there are no osteoderms on top or embedded between the ribs. This was confirmed by embryological studies, which track the development of the turtle's carapace from the developmental changes in the back ribs; no osteoderms are involved.

From the turtle on a half shell, we can trace the ancestry of turtles even further back to reptiles that have only a few turtle-like features. One of these

Figure 12.5 ▶

Odontochelys: (A) the best of the known specimens of this genus, showing an incomplete carapace on its back (*left*) but a complete plastron on its belly (*right*); (B) a reconstruction of its appearance in life. ([A] Courtesy of Li Chun; [B] courtesy of Nobumichi Tamura)

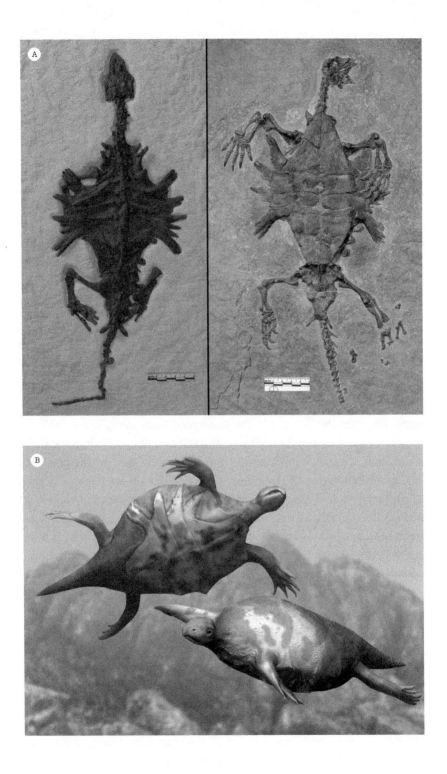

is *Eunotosaurus*, from the Middle Permian beds (about 270 million years old) of South Africa (figure 12.6A). It looked mostly like a large, fat lizard, except for some key features of the skeleton. The most striking of these is the greatly expanded broad flat back ribs that almost connect with each other to form a complete shell on the back. And in 2015, another primitive fossil was announced by my friend Hans-Dieter Sues of the Smithsonian. Named *Pappochelys* (grandfather turtle), it not only has the broad ribs on its back like *Eunotosaurus* but also broad flattened belly ribs (gastralia) that would eventually fuse into the plastron or belly plate of more advanced turtles. Thus we have a very nice transition from the reptile *Eunotosaurus*, with only flattened back ribs, to *Pappochelys* with flattened back and belly ribs, to *Odontochelys*, with a belly shield but only flattened ribs on the back, to *Proganochelys*, with a complete (but primitive) turtle shell, but still retaining some teeth, and unlike modern turtles in having not yet developed the ability to retract its head (figure 12.6B). One could not ask for a better transition for a distinct type of animal than the fossils that most would not guess would lead to turtles.

For a final example, let's look at snakes. Snakes are a very distinctive group of animals, with their highly specialized legless bodies adapted to all sorts of niches, from burrowing to tree climbing. Snakes are also highly symbolic, from the Serpent in the Garden of Eden myth in the Bible to their symbolic use in many cultures. In ancient Egypt, the cobra adorned the crown of the Pharaoh, and in Greek myths the Gorgon Medusa had snakes for hair. Herakles (Hercules) had to kill the Lernean Hydra by cutting off its nine snake heads, each of which grew a new head as soon as it was cut. The Greeks also revered snakes in medicine, so the symbol of healing (the Caduceus) was a rod with a snake intertwined around it. Snakes are worshipped in the Hindu and Buddhist religions; the neck of the Hindu god Shiva is wrapped by cobras, and Vishnu was depicted as sleeping on a seven-headed snake or within a snake's coils. Snakes were also an important part of Mesoamerican mythology and religion. The Chinese have long revered snakes and have eaten them in their cuisine as a delicacy as well. One of the twelve signs of the Chinese zodiac is the "Snake."

Snakes also evoke strong feelings in humans even though most are no more dangerous than a bunny or lizard. For many people, their cold stare with unblinking eyelids, their flicking tongues, or the unnerving way they slither and move about without legs is unsettling. However, the fact that some are venomous is the source of most fears of snakes. In some parts of the world, such as here in the United States, only a few snakes (primarily rattlesnakes) are venomous,

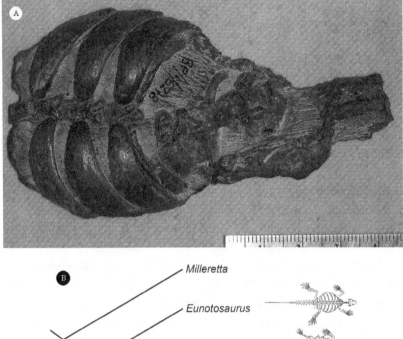

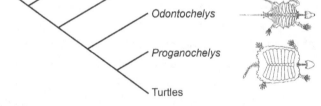

Figure 12.6 ▲

Eunotosaurus was a primitive Permian reptile with flared ribs that suggest the earliest stages of turtle shell evolution. (*A*) Partial specimen shows the distinctive flange-shaped ribs, which make a partial shell. (*B*) Relationships of *Eunotosaurus* and other primitive turtles, showing the transition from primitive reptiles to turtles. ([*A*] Courtesy of B. Rubidge; [*B*] redrawn from several sources)

but people will kill all snakes on sight, even though they are extremely useful in reducing the problems of rats or mice and other pests. However, in Australia the 10 most common snakes are extremely venomous, and many of those in Africa are venomous as well. Thus the primal fear of snakes (ophidiophobia) is deeply rooted in our primate brain and often cannot be overcome, no matter how harmless a typical bull snake or gopher snake really is.

Snakes are diverse, with about 2,900 species alive today in 29 families and dozens of genera. They live on every continent and in every habitat except areas that are extremely cold, such as the Arctic or Antarctic, or on islands so isolated that they never receive snakes from the mainland (New Zealand, Ireland, Iceland, Hawaii, and most of the South Pacific). Ireland, by the way, lacks snakes not because St. Patrick drove them out but because it was under ice during the last Ice Age. When the ice retreated and sea level rose, Ireland became an island, so snakes never had an opportunity to reach it.

Snakes are so diverse because they can adapt to a variety of habitats, from trees to the ground to underground burrows dug by their prey. Lacking limbs, they move into areas (like burrows) by sliding along, and they can wrap their bodies around limbs (as in tree-climbing snakes) or even progress across hot desert sand using their unique method of travel known as sidewinding. Being limbless has its limitations: They cannot grab prey so must coil around it to restrain it; in one group, the constrictors, they squeeze the prey until it cannot inhale any more and dies. Without limbs or powerful crushing jaws such as those of crocodiles, they cannot rip prey apart but must swallow it whole. Snakes have highly flexible and stretchable skulls, with most bones reduced to tiny splints hinged to the rest of the skull and held together with stretchy ligaments, so they can expand and wrap their entire head around a prey item wider than they are, and slowly swallow it.

The consequence of their unique anatomy is that snake fossils are extremely rare. Their delicate skull bones break up easily, so all that remains are isolated vertebrae and delicate ribs of the backbone, none of which fossilize easily. Most fossil snakes are known only from distinctive vertebrae and not much else. One would think that no fossil record of how snakes evolved from something else could be found, let alone a fossil that is in transition from having functional limbs to losing its limbs. Despite the odds, there is a good record of that transition.

We already have a number of closely related lizard groups, especially the Varanidae or monitor lizards (such as the Komodo dragon, the goanna of Australia, and many others), which have many snake-like features in their skulls and skeletons but still have four strong limbs. And in the past 20 years, a whole host of snake fossils have appeared that show how those limbs slowly became tinier and tinier, until they were lost completely. The least snake-like of these fossils is *Adriosaurus microbrachis* from the middle Cretaceous (95 million years old) rocks of Slovenia (figure 12.7E, figure 12.7F). It was a marine lizard-like fossil with an extremely long slender snake-like

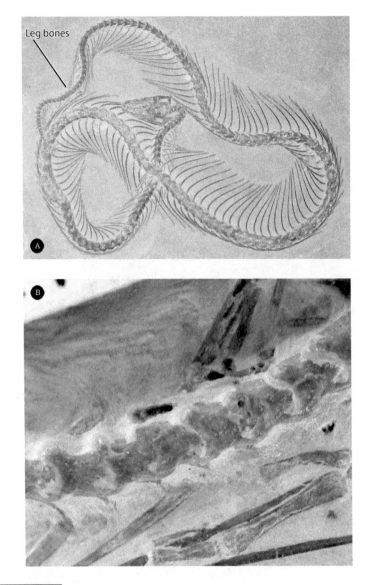

Leg bones

Figure 12.7 ▲ ▶

A number of transitional snake fossils have vestigial legs and hip bones and are known from the Cretaceous. (A) *Eupodophis descouensi*, with tiny vestigial hind legs. (B) Detail of the leg bones in the same specimen. (C) The complete articulated skeleton of the Cretaceous snake with legs known as *Haasiophis*. The large orange cubes are cork spacers to prevent the fossil from being damaged when it is turned upside down. (D) Detail of the hip region, showing the vestigial hind limbs. (E–F) The transitional fossil *Adriosaurus*, which had functional hind limbs but vestigial forelimbs and a long snake-like body. ([A–B] Courtesy of Michael W. Caldwell; [C–D] courtesy of Michael J. Polcyn; [E–F] after Alessandro Palci and Michael W. Caldwell, "Vestigial Forelimbs and Axial Elongation in a 95 Million-Year-Old Non-Snake Squamate," *Journal of Vertebrate Paleontology* 27, no. 1 [2007]: 1–7)

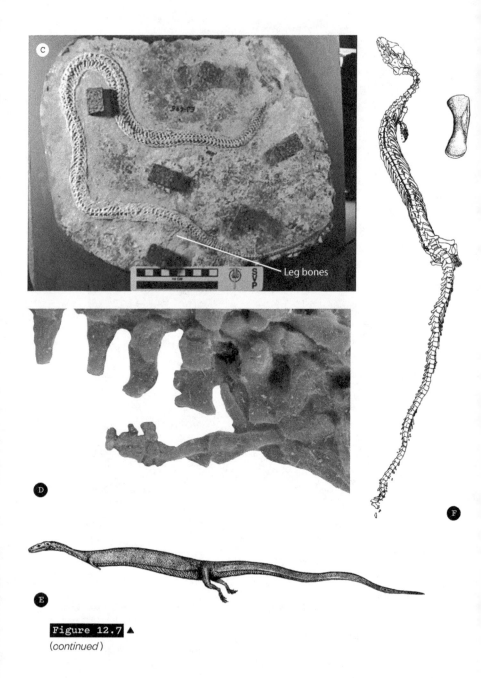

Leg bones

Figure 12.7 ▲
(*continued*)

body—but it had only tiny limbs in front and in back, barely able to help with propulsion. *Najash rionegrina* from the middle Cretaceous (about 90 million years old) of Argentina still has hip bones, and vestigial hind limbs attached to the hips retain most of the elements (thighbone, shin bone, and some foot elements), but no forelimbs.

The Upper Cretaceous marine rocks of the Middle East (especially Lebanon and Israel) have yielded a wealth of complete skeletons of transitional snake fossils. *Haasiophis terrasanctus* is about 94 million years old, with a skull and vertebrae like other primitive snakes, but no forelimbs, and extremely tiny hip bones and thighbones that do not attach to one another, so they are truly vestigial and no longer could function (figure 12.7C; figure 12.7D). *Pachyrachis* also has tiny vestigial hind limbs, no forelimbs, and thick dense bone in its ribs and vertebrae that would help with diving. *Eupodophis descouensi* was found in rocks about 92 million years old in Lebanon, and they had even tinier vestigial hind limbs (figure 12.7A; figure 12.7B).

Snakes with legs? The fossil record provides them in abundance, from *Adriosaurus*, the last snake relative to have both tiny forelimbs and hind limbs; to *Najash*, without forelimbs but with some function to its hind limbs; to *Haasiophis, Pachyrachis*, and *Eupodophis*, which have tiny truly vestigial hind limbs that had no function but were mute witnesses to the days when all snakes had legs.

Clinching this fossil record is a surprising fact about living snakes: Some of them (mostly the boas and their relatives) still have tiny, nonfunctional remnants of their hip bones and thighbones buried deep in their bodies (see figure 10.1A). In a few species, these tiny thighbones project out as a scaly "spur" on the side of the body, but none of these bones has any real function now other than being proof that snakes evolved from ancestors with legs.

Finally, losing legs is not that big of a deal. It has happened in many different groups of four-legged animals, all independently evolved. Examples of leg loss in the four-legged tetrapods include not only the snakes but an entire group of living reptiles called the amphisbaenians, as well as several different groups of legless lizards, including some skinks, the Australian flap-footed lizards, "slow worms," "glass lizards," and several others. Among amphibians, an entire group (the caecilians or apodans) developed worm-like bodies, and a group called the sirens have only stunted forelimbs and no hind limbs. In addition, at least two extinct groups of amphibians, the aistopods and lysorophids, became limbless as well. Nearly every one of these examples is a burrowing animal, and the loss of limbs appears to aid in digging through the ground or soft mud. There's a simple reason losing all the limbs is so easy. The development of the limb buds and eventually the limbs is controlled by a specific set of Hox genes and Tbx genes; all it takes is for those genes to shut off the commands to develop limbs and the limbs will vanish.

Next time you see a turtle, a snake, or a frog, think about how they are no longer isolated and unrelated groups in the animal kingdom. The fossil record of how they originated from very different looking ancestors has been found, and all the transitional fossils we need are in our possession. They are no longer part of "a chain with missing links."

FOR FURTHER READING

Anderson, Jason S., Robert R. Reisz, Diane Scott, Nadia B. Fröbisch, and Stuart S. Sumida. "A Stem Batrachian from the Early Permian of Texas and the Origin of Frogs and Salamanders." *Nature* 453, no. 7194 (2008): 515–518.

Caldwell, Michael W., and Michael S. Y. Lee. "A Snake with Legs from the Marine Cretaceous of the Middle East." *Nature* 386 (1997): 705–709.

Li, Chun, Xiao-Chun Wu, Olivier Rieppel, Li-Ting Wang, and Li-Jun Zhao. "An Ancestral Turtle from the Late Triassic of Southwestern China." *Nature* 456 (2008): 497–501.

Gauther, J. A., A. G. Kluge, and T. Rowe. "The Early Evolution of the Amniota." In *The Phylogeny and Classification of the Tetrapods*, vol. 1, *Amphibians, Reptiles, Birds*, ed. Michael J. Benton, 103–155. Oxford: Clarendon Press, 1988.

Held, Lewis I., Jr. *How the Snake Lost Its Legs: Curious Tales from the Frontiers of Evo-Devo*. Cambridge: Cambridge University Press, 2014.

Joyce, Walter G. "Phylogenetic Relationships of Mesozoic Turtles." *Bulletin of the Peabody Museum of Natural History* 48, no. 1 (2007): 3–102.

Lyson, Tyler R., Gabe S. Bever, Bhart-Anjan S. Bhullar, Walter G. Joyce, and Jacques A. Gauthier. "Transitional Fossils and the Origin of Turtles." *Biology Letters* 6, no. 6 (2010): 830–833.

Prothero, Donald. *Evolution: What the Fossils Say and Why It Matters*. 2nd ed. New York: Columbia University Press, 2017.

Rieppel, Olivier. "A Review of the Origin of Snakes." *Evolutionary Biology* 25 (1988): 37–130.

Rieppel, Olivier, Hussam Zaher, Eltan Tchernov, and Michael J. Polcyn. "The Anatomy and Relationships of *Haasiophis terrasanctus*, a Fossil Snake with Well-Developed Hind Limbs from the Mid-Cretaceous of the Middle East." *Journal of Paleontology* 77, no. 3 (2003): 536–558.

Rieppel, Olivier, and Michael deBraga. "Turtles as Diapsid Reptiles." *Nature* 384 no. 6608 (1996): 453–455.

Schoch, Rainer R., and Hans-Dieter Sues. "A Middle Triassic Stem-Turtle and the Evolution of the Turtle Body Plan." *Nature* 523, no. 7562 (2015): 584–587.

BIRDS WITH TEETH

And if the whole hindquarters, from the ilium to the toes, of a half-hatched chick could be suddenly enlarged, ossified, and fossilised as they are, they would furnish us with the last step of the transition between Birds and Reptiles; for there would be nothing in their characters to prevent us from referring them to the Dinosauria.

—THOMAS HENRY HUXLEY, *FURTHER EVIDENCE OF THE AFFINITY*
BETWEEN DINOSAURIAN REPTILES AND BIRDS (1870)

In 1892, a severe drought struck the western United States. There were also problems with water distribution, and the settlers were howling for Congress to get involved. They were particularly incensed at the U.S. Geological Survey (USGS). Since 1881 it had been led by the one-armed Civil War veteran John Wesley Powell, famous for leading the first expedition down the Colorado River through the Grand Canyon in 1869. Many were angry because Powell's 1878 *Report on the Lands of the Arid Regions of the United States* warned that water would always be a problem in the arid west and that the region did not have enough water for large-scale farming and settlement. In an 1883 conference on water rights, Powell prophetically warned: "Gentlemen, you are piling up a heritage of conflict and litigation over water rights, for there is not sufficient water to supply the land." However, the powerful railroad companies were making big money selling their excess land to settlers (over 183 million acres that they had been granted when they built the first transcontinental railroad). They wanted to continue to encourage settlement, and they bribed congressmen to advance their interests. Congress responded to the pressures of the drought by blaming the messenger rather than by listening to his warnings, which were proved true during the Dust Bowl years of the 1930s.

Congressional investigators scrutinized every part of the USGS and how it had spent its money, trying to find evidence of corruption or waste (common in many political and government organizations at that time). Among the discoveries was that Professor Othniel Charles Marsh of Yale University, also serving as the USGS chief paleontologist, had published a huge glossy monograph in 1880 on some spectacular complete fossil bird skeletons from the marine chalk beds of western Kansas. They included two archaic Cretaceous birds, the loon-like *Hesperornis* (figure 13.1A) and the tern-like *Ichythornis* (figure 13.1B), which looked similar to their living counterparts—except they had teeth. Marsh had named a group for these ancestral toothed birds the Odontornithes ("toothed birds" in Greek). Fundamentalist Alabama congressman Hilary Herbert got wind of this

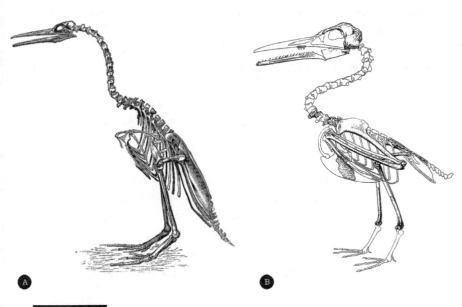

Figure 13.1 ▲

The skeletons of the (*A*) loon-like *Hesperornis* and the (*B*) tern-like *Ichthyornis*, both toothed birds found in the Cretaceous marine chalk beds of western Kansas. (From Othniel Charles Marsh, *Odontornithes: A Monograph on the Extinct Toothed Birds of North America; with Thirty-Four Plates and Forty Woodcuts* [Washington, D.C.: Government Printing Office, 1880]) and was incensed. To him the idea that the USGS was publishing books about biblical impossibilities like toothed birds was a bigger outrage than ordinary corruption and misuse of tax dollars. On the floor of Congress, he ranted, "Birds with teeth! That's where your hard-earned money goes, folks—on some professor's silly birds with teeth." Shortly thereafter, Congress severely cut the budget of the USGS, especially their paleontological research. Powell sent Marsh a blunt telegraph: "Appropriation cut off. Submit your resignation at once." A year later Powell himself was forced to resign. Ironically, he was replaced by the famous trilobite paleontologist Charles Doolittle Walcott, who was an even bigger supporter of evolution than Powell had been.

But the discovery of toothed birds was not new. In fact, the first such example was the famous transitional fossil *Archaeopteryx*, first named and described in 1861, just two years after Darwin's book was published. The first skeleton of this legendary fossil had been described by Richard Owen in 1863 when the British Museum bought it and imported it from Germany. Even though Owen was a bitter critic of Darwin and his ideas, and did his best to minimize the dinosaurian features of *Archaeopteryx*, its importance was clear to Darwin and his supporter Thomas Henry Huxley. By the fourth edition of *On the Origin of Species*, Darwin would brag that at one time some scientists argued

> that the whole class of birds came suddenly into existence during the Eocene period [54–34 million years ago, as we now date it]; but now we know, on the authority of Professor Owen, that a bird certainly lived during the deposition of the Upper Greensand [Late Early Cretaceous in modern terminology, about 100 million years ago; this specimen turned out to be a pterosaur]; and still more recently, that strange bird, the *Archaeopteryx*, with a long lizard-like tail, bearing a pair of feathers on each joint, and with its wings furnished with two free claws, has been discovered in the oolitic slates of Solnhofen. Hardly any recent discovery shows more forcibly than this how little we as yet know of the former inhabitants of the world.

Since that time at least 11 or 12 more specimens of *Archaeopteryx* have been found. The best and most famous one (figure 13.2), now in the Museum für Naturkunde in Berlin, shows the complete skeleton in a naturalistic death pose, and the skull and teeth are much better preserved than in the

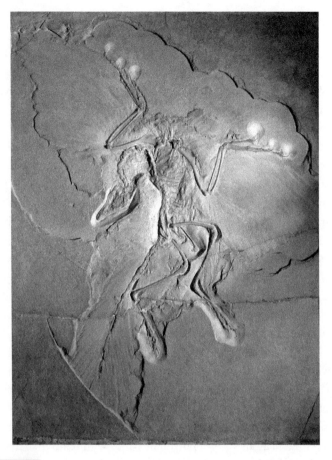

Figure 13.2 ▲

The most famous skeleton of *Archaeopteryx*, found in the Upper Jurassic Solnhofen Limestone in Bavaria in 1877 and now held in the Museum für Naturkunde, Berlin. It is the most complete specimen of the 12 or 13 that have been discovered. (Courtesy of Wikimedia Commons)

London specimen. Looking at *Archaeopteryx*, Huxley realized how much it showed that birds were simply modified dinosaurs. For one thing, he had recently described the skeleton of the small dinosaur *Compsognathus*, from the same Jurassic Solnhofen Limestone quarries as *Archaeopteryx* (this dinosaur became famous as the "Compies" in the *Jurassic Park/Jurassic World* movies and books). Huxley was struck by the similarities in their skeletons even though one had preserved feathers and was called a bird and the other did not and was just a dinosaur. In fact, another specimen of *Archaeopteryx* was first misidentified as *Compsognathus*; only later did John Ostrom of Yale University realize that it was actually *Archaeopteryx*.

But the list of dinosaurian features in *Archaeopteryx* goes far beyond the teeth (figure 13.3A). *Archaeopteryx* has a long bony tail, a dinosaurian feature found in no living bird, whose tail bones are all fused into a tiny reduced nub of bones called the pygostyle or "parson's nose." The skull of *Archaeopteryx* has the same arrangement of holes in the side found in dinosaurs and is especially similar to that of the predatory dinosaurs like *Velociraptor*; it is very different from the highly modified skulls of modern birds. The vertebrae are also like those of dinosaurs and are not arranged in the flexible configuration seen in modern birds. The hip bone is intermediate between that of dinosaurs and birds, as is the strap-like shoulder blade. *Archaeopteryx* has gastralia, or belly ribs, found in many predatory dinosaurs but not found in modern birds.

Its most striking feature is the configuration of the hand and wrist. *Archaeopteryx* had long claws like those of predatory dinosaurs, not birds, and it had fully functional hands with fingers 1–3 (thumb, index, middle fingers) like all the theropod dinosaurs, not the fused hand bones found in a bird wing, called the carpometacarpus and the alula. These bones form the small triangular pointed bit of bone at the end of the chicken wings you order for a meal, which you never eat because there is no meat (muscle) on them. Instead of fingers supporting their wings (as in bats), birds have almost no fingers at all and support their wings with feather shafts. In the wrist, birds and *Velociraptor* have a unique configuration of wrist bones fused into a half-moon shape, called the semilunate carpal (figure 13.3B). With this kind of wrist configuration, *Velociraptor* and its relatives can strike downward and forward quickly with their hands—but they cannot rotate their palms inward (which is commonly seen on incorrect reconstructions of dinosaurs). In other words, they could not catch a basketball between their hands, but they could rotate their palms down to dribble. That rapid downward and forward snap of the wrist is the same motion that you see in the downstroke of the wing of a bird during flight—and it's all due to the semilunate carpal in the wrist.

The clincher is found in the hind legs of *Archaeopteryx*. They have a unique ankle configuration called the mesotarsal joint, found only in dinosaurs, birds, and their close relatives, the pterosaurs (figure 13.3C). Most

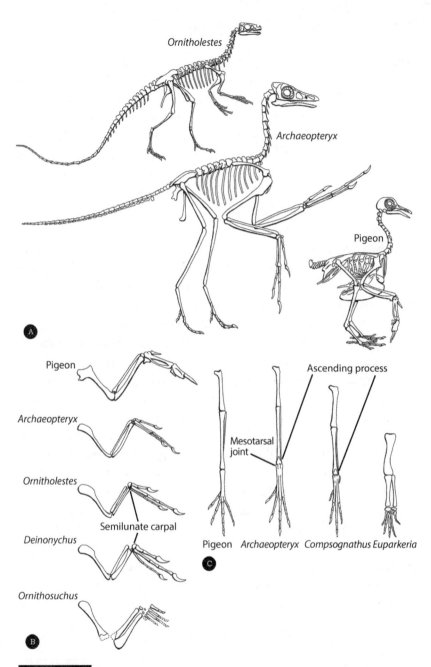

Figure 13.3 ▲

(A) Comparison of the anatomical features of *Archaeopteryx* with a more advanced bird, and with a small theropod dinosaur like *Ornitholestes*. (B) Anatomy of the forelimb of birds and dinosaurs, showing the semilunate carpal, or wrist bone. (C) Anatomy of the hind leg of birds and dinosaurs, showing the mesotarsal joint and ascending process of the astragalus. (Drawing by Carl Buell; from Donald Prothero, *Evolution: What the Fossils Say and Why It Matters*, 2nd ed. [New York: Columbia University Press, 2017]) vertebrate ankles (including yours) are made of a series of rows of ankle bones, and they have a hinge between the tibia (shin bone) and the first row of ankle bones (calcaneum and astragalus). However, in birds, dinosaurs, and pterosaurs, the ankle hinge is between the first and second row of ankle bones, so the astragalus and calcaneum actually fuse to the end of the tibia. Next time you eat a chicken or turkey drumstick, notice the little cap of cartilage on the "handle" end of the drumstick. These are the first row of ankle bones, a dinosaurian feature found in every bird known. In addition, all birds and dinosaurs have a bony spur that sticks up from the astragalus along the front of the tibia, another unique dinosaurian feature. The toes and feet of *Archaeopteryx* are like those of dinosaurs rather than most birds, and its big toe was not opposable nor capable of grasping branches, unlike those of most perching birds today. *Archaeopteryx* even had one enlarged toe claw similar to the slashing claws seen in *Velociraptor* and its kin.

In most respects, *Archaeopteryx* is just a dinosaur with feathers. Only the reversal of the big toe to point backward, the lack of steak-knife serrations on the edges of the teeth, the asymmetrical flight feathers, and the relatively large arms distinguish it from dinosaurs like *Velociraptor*. In the past few decades, hundreds of amazing discoveries (especially from the Lower Cretaceous beds of Liaoning Province, China) have produced all sorts of primitive toothed birds, beautifully preserved with their feathers intact, and some with the original coloration visible, as well as rare specimens with stomach contents or internal organs preserved. In addition, these same beds produce a spectrum of nonbird dinosaurs, showing that feathers are found in all groups of dinosaurs (and since 2018 we have found specimens of pterosaurs that show they had feathers too). Most of these feathered nonbird dinosaurs do not have flight feathers; the feathers are there for their original purpose, insulation. This is just as it is for modern birds, who use only a small percentage of their wing feathers and tail feathers for flight. Most of their feathers are body feathers such as down, which hold in their body heat.

Archaeopteryx was the first "missing link" to show the evolution of birds from dinosaurs, but there is now a huge flock of intermediate birds more advanced than *Archaeopteryx* that provide the steps in the transition to living birds (figure 13.4). *Rahonavis*, from the Cretaceous of Madagascar, also had a sickle-like claw on the hind feet, a long bony tail, teeth, and several other dinosaurian features. But like more advanced birds, its hips are fused

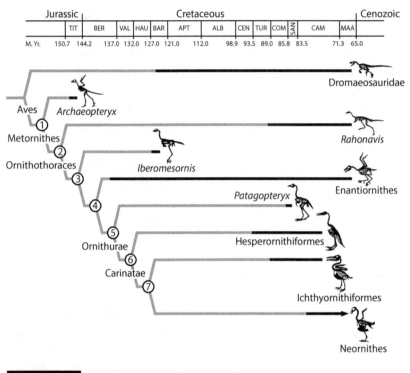

Jurassic					Cretaceous									Cenozoic
	TIT	BER	VAL	HAU	BAR	APT	ALB	CEN	TUR	COM	SAN	CAM	MAA	
M. Yr.	150.7	144.2	137.0	132.0	127.0	121.0	112.0	98.9	93.5	89.0	85.8	83.5	71.3	65.0

Figure 13.4 ▲

The family tree of Mesozoic birds, emphasizing some of the recent fossil discoveries. (Courtesy of Luis M. Chiappe)

to the lower back vertebrae to form a synsacrum, and it has holes in its vertebrae for the air sacs found in living birds. It even has quill knobs (bumps where the feathers attached to the bone) on its arms and fingers, showing it had robust flight feathers and was probably a better flier than *Archaeopteryx*. One of its most bird-like features is the fibula, the tiny bone that runs parallel to the shin bone no longer reached all the way down to the ankle, as in *Archaeopteryx*, but tapered down into nothing. If you've ever eaten a chicken or turkey drumstick, you will find this tiny toothpick of a bone that does not connect to the ankle but is embedded in the muscles of the leg.

One step up from *Rahonavis* is the Early Cretaceous Chinese bird *Confuciusornis*. It is just slightly more advanced; the many separate tail bones of *Archaeopteryx* and primitive birds and dinosaurs have been reduced to a pygostyle, which is almost as small as that of modern birds. The next step upward is an entire group of the dominant Cretaceous birds, or Enantiornithes (opposite birds). They are more advanced than *Archaeopteryx*, *Rahonavis*, or *Confuciusornis*, having fewer trunk vertebrae, a flexible wishbone, fusion of the hand bones to form the carpometacarpus and alula, and a shoulder joint much better for flying (see figure 13.4).

Then we come to even more advanced Cretaceous birds such as *Vorona* from Madagascar, *Patagopteryx* from Argentina, and the well-known aquatic birds *Hesperornis* and *Ichthyornis* from the chalk beds of Kansas (see figure 13.1). They resemble dinosaurs less and have many more bird-like features, including loss of the belly ribs, reorientation of the pubic bone backward parallel to the ischium to form the classic "bird hip," fewer trunk vertebrae, and additional features of the hand and shoulder. *Ichthyornis* even had a strong keel on the breastbone for powerful flight muscles—and they *still* had teeth!

Finally, we come to the earliest members of the living class Aves, or modern birds. Many anatomical features in Aves are not found in their ancestors, including the loss of the teeth (finally) and the complete fusion of the leg bones and first row of ankle bones to form what is called a tarsometatarsus. The transitional sequence from birds to dinosaurs is now one of the most completely documented in the fossil record, but where you draw the line between bird and dinosaur is a nearly impossible task because we now think of birds as being a subgroup of dinosaurs—so the dinosaurs are not extinct.

Have birds completely lost their teeth? One would think so because teeth are never seen on living birds, which have a horny beak instead. This idea is reflected in the saying "as scarce as hen's teeth" (as in so scarce that they are never found). But a pioneering experiment in embryology and genetics in 1980 by E. J. Kollar and C. Fisher had a surprising outcome. They grafted the mouth epithelium of a lab mouse into the mouth of a developing chick's beak. They let the chick develop and were stunned to find that somehow it had developed teeth again! But these were not mouse teeth at all—they were the simple conical teeth of predatory dinosaurs and the Cretaceous birds that still had teeth. Apparently, birds still have the information to make dinosaurian teeth in their genes, but this has been suppressed by their regulatory genes and is never expressed—except when tampered with in scientific experiments.

A lot of old genes are still present in animals, such as our genes for a tail (chapter 5) or the genes for the side toes of a horse (chapter 14). These are manifested as "evolutionary throwbacks" or atavisms, illustrating that the instructions for making a feature that is no longer needed may not be eliminated during evolution. Instead, the genes for it may just be silenced. Since the famous Kollar and Fisher experiment, scientists have found lots of other dinosaurian genes that were repressed in birds but can be expressed if the shut-off command is eliminated. One study managed to manipulate the chick genome so the bird developed a long bony dinosaurian tail, like that in *Archaeopteryx*, not the short stubby pygostyle of modern birds. Another experiment tampered with the chick genes so their feet look dinosaurian, not bird-like. Even more amazing is genetic manipulation of the genes for the mouth of a bird, which produced a dinosaurian mouth with teeth rather than the toothless beak of modern birds.

In short, we have an excellent fossil record that shows the transition of *Velociraptor* to *Archaeopteryx* to modern birds. It is an established fact that birds are descended from small predatory dinosaurs, and it's proper to say that "birds are living dinosaurs." In addition, we know what genetic switches were used to transform the bird skeleton during this evolutionary history. The next time you look up and hear a song or see a feathered animal flying by, enjoy the dinosaurs that continue to thrive on our planet.

FOR FURTHER READING

Chiappe, Luis M. "The First 85 Million Years of Avian Evolution." *Nature* 378 (1995): 349–355.

Chiappe, Luis M., and Gareth J. Dyke. "The Mesozoic Radiation of Birds." *Annual Review of Ecology and Systematics* 33 (2002): 91–124.

Chiappe, Luis M., and Lawrence M. Witmer, eds. *Mesozoic Birds: Above the Heads of Dinosaurs.* Berkeley: University of California Press, 2002.

Currie, Philip J., Eva B. Koppelhus, Martin A. Shugar, and Joanna L. Wright, eds. *Feathered Dragons: Studies on the Transition from Dinosaurs to Birds.* Bloomington: Indiana University Press, 2004.

Dingus, Lowell, and Timothy Rowe. *The Mistaken Extinction.* New York: W. H. Freeman, 1997.

Gauthier, Jacques A. "Saurischian Monophyly and the Origin of Birds." In *The Origin of Birds and the Evolution of Flight,* ed. Kevin Padian, 1–56. San Francisco: California Academy of Sciences, 1986.

Gauthier, Jacques A., and Lawrence F. Gall, eds. *New Perspectives on the Origin and Early Evolution of Birds.* New Haven, Conn.: Yale University Press, 2001.

Kollar, E. J., and C. Fisher. "Tooth Induction in Chick Epithelium: Expression of Quiescent Genes for Enamel Synthesis." *Science* 207, no. 4434 (1980): 993–995.

Norell, Mark A. *Unearthing Dragons: The Great Feathered Dinosaur Discoveries.* New York: Pi Press, 2005.

Ostrom, John H. "*Archaeopteryx* and the Origin of Birds." *Biological Journal of the Linnaean Society* 8, no. 2 (1976): 91–182.

——. "*Archaeopteryx* and the Origin of Flight." *Quarterly Review of Biology* 49, no. 1 (1974): 27–47.

Padian, Kevin, and Luis M. Chiappe. "The Origin of Birds and Their Flight." *Scientific American* 278, no. 2 (February 1998): 28–37.

Prum, Richard O., and Alan H. Brush. "Which Came First, the Feather or the Bird?" *Scientific American* 288 (March 2004): 84–93.

Shipman, Pat. *Taking Wing:* Archaeopteryx *and the Evolution of Bird Flight.* New York: Simon & Schuster, 1988.

A HORSE! A HORSE! MY KINGDOM FOR A HORSE!

The geologic record of the ancestry of the horse is one of the classic examples of evolution.

—WILLIAM DILLER MATTHEW, 1926

One of the first evolutionary sequences to be discovered was that of horses, and by the 1920s the classic diagram of horse evolution (figure 14.1) was popularized and established in nearly every textbook example of evolution. Ironically, that familiar iconic image is grossly out of date. Thousands of horse fossils have been uncovered in the past 95 years, and they provide a much better, more detailed—and very different—picture of their evolution. But let's save that story for the end of this tale.

The story begins in England in 1839, during the heyday of the English mania for natural history collecting. William Richardson, Esq., MA, FGS, a gentleman collector, was prospecting on the coast of Kent in the lower Eocene beds of the famous London Clay. He was looking for "strong expectations for the evidence of some form of animal life, whether of beast or bird, destined to be sustained by so rich a provision." He was lucky this time because he found the front half of a tiny skull (figure 14.2), as well as parts of a fossil bird.

This little skull was given to Richard Owen at the British Museum, who was the foremost zoologist and paleontologist in England at that time. Indeed, Owen was famous for describing the first dinosaurs and even

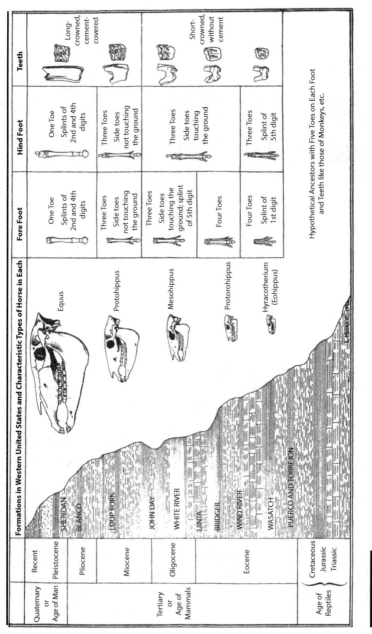

Figure 14.1 ▲

William Diller Matthew's 1925 diagram of the evolution of the horse, showing the general trends in larger body size, longer snouts, higher-crowned cheek teeth, longer legs, and reduction of side toes that the few fossils known at the time demonstrated. At that time, the few available fossils suggested a single lineage with a linear trend of change. (From William Diller Matthew, "The Evolution of the Horse: A Record and Its Interpretation," *Quarterly Review of Biology* 1, no. 2 [April 1926]: 139–185)

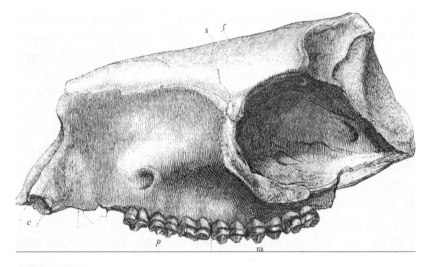

Figure 14.2 ▲

Richard Owen's original type skull of *Hyracotherium leporinum* is in the collections of the Natural History Museum in London. (From Richard Owen, "Description of the Fossil Remains of a Mammal (*Hyracotherium leporinum*) and of a Bird (*Lithornis vulturinus*) from the London Clay," *Transactions of the Geological Society of London*, Series 2, 6 [1841]: 203–208)

coined the word "dinosaur." He had worked on many famous fossils, including giant extinct ground sloths and Darwin's strange South American fossils from the voyage of the HMS *Beagle*. The little skull, with large eyes and a short snout, looked more "like that of a Hare or other timid Rodentia." However, the low rectangular teeth with small cusps were clearly those of a primitive hoofed mammal that most closely resembled the extinct *Choeropotamus* (a very primitive even-toed hoofed mammal from the same beds) among the handful of fossil mammals known at that time. Owen correctly realized that the peculiar arrangement of cusps and ridges was even more similar to the living hyrax, so he named the little skull *Hyracotherium leporinum*, or "rabbit-like hyrax beast."

A few years later, Owen was describing some more Eocene mammals, this time from the Isle of Wight on the south coast of England. After describing the fossils, he discussed some ideas first suggested by Cuvier in 1817 and H. M. D. de Blainville in 1816. Both of these French zoologists had argued that hoofed mammals could be classified by the number of their toes.

Some have an even numbers of toes (two or four) and were in one group, the Artiodactyla; others had odd numbers (three or one) and belonged to a different group. In 1848, Owen adopted de Blainville's association of horses, rhinos, tapirs, and hyraxes, and coined the name Perissodactyla for them. He did not put his little *Hyracotherium* in this group, however, because he had nothing but the skull and no foot bones yet.

Ironically, *Hyracotherium* turned out to be the most ancient and primitive perissodactyl fossil then known, but it was not recognized as such until the 1870s. Instead, attention focused elsewhere. In 1859, Charles Darwin published *On the Origin of Species*, and all of biology was turned upside down. As the debates in scientific circles became more and more bitter, critics pointed to the shortage of good examples of sequences of fossils that led to living animals. There were spectacular examples of transitional fossils, such as the half-bird half-dinosaur *Archaeopteryx*, but the record of fossil mammal evolution was very incomplete in Europe. Nevertheless, some patterns were beginning to emerge. In 1872, Darwin's chief defender, Thomas Henry Huxley, pointed out that three fossil mammals from Europe, if placed in order of their age, formed a sequence leading to the modern horse, genus *Equus*. There was the bizarre Eocene tapir-like *Palaeotherium* from the middle Eocene gypsum beds of Montmarte, in northern Paris, France, described by Cuvier; the early Miocene browsing horse *Anchitherium*; and the late Miocene grazing horse *Hipparion*. The next year Russian paleontologist Vladimir Kovalesky studied the same fossils, and he was even more certain that they represented the ancestral sequence of the modern horse. Both Huxley and Kovalesky realized that the sequence was patchy and incomplete. There were just four fossil horses, with large gaps in between. Nevertheless, they correctly concluded that horses arose from a more tapir-like animal with three toes and very low-crowned teeth.

Unfortunately for these European scientists, their sequence was not representative of the main line of horse evolution because that story took place elsewhere. Those European horse relatives were immigrant side branches from North America, where most of the history of the horse took place. Horse fossils from the Big Badlands of South Dakota were first described by Joseph Leidy in 1850. He first referred to these fossils as *Palaeotherium* from Europe, and then assigned them to the horse *Anchitherium*, not realizing that they were new forms unknown in Europe. By 1869 he had described quite a few fossil horses, which had been found all over western North America.

Yale paleontologist Othniel Charles Marsh took the next step. In 1871 and 1872, he began to find Eocene horses in the Rocky Mountains. He also found other horses that filled in some of the gaps, and he began to work out the changes in their teeth, limbs, and feet (figure 14.3A) from his more complete skeletons, which neither Leidy nor any European paleontologist had seen. By 1874, Marsh boasted that "the line of descent appears to be direct, and the remains now known supply every important form." In 1873, his bitter rival Edward Drinker Cope described specimens he had received from the early Eocene beds in Wyoming as *Eohippus* (dawn horse). A few years later Cope realized that it was the American equivalent of *Hyracotherium* and placed *Eohippus* at the base of horse evolution.

Thomas Henry Huxley sailed to America during its centennial year in 1876 to give lectures on topics in natural history. He planned to give a learned discourse on the evolution of the horse in Europe, based on the work that he and Kovalesky had done. However, he spent two days with Marsh in the collections at Yale and found that Marsh's evidence was convincing. As his son and biographer Leonard Huxley wrote,

> At each inquiry, whether he had a specimen to illustrate such and such a point or to exemplify transition from earlier and less specialized forms to later and more specialized ones, Professor Marsh simply turned to his assistant and bid him fetch box number so-and-so, until Huxley turn upon him and said, "I believe you are a magician; whatever I want, you just conjured it up."

As Marsh later recalled this encounter:

> He then informed me that this was new to him, and that my facts demonstrated the evolution of the horse beyond question, and for the first time indicated the direct line of an existing animal. With the generosity of true greatness, he gave up his own opinions in the face of new truth and took my conclusions.

Huxley later wrote to Marsh, "The more I think of it, the more clear it is that your great work is settlement of the pedigree of the horse."

Huxley then rewrote his lecture, using diagrams supplied by Marsh to recant his old ideas and present the new evidence. Horses had evolved in America, and his European examples were occasional immigrants from the New World. Huxley was overjoyed with Marsh's evidence because it

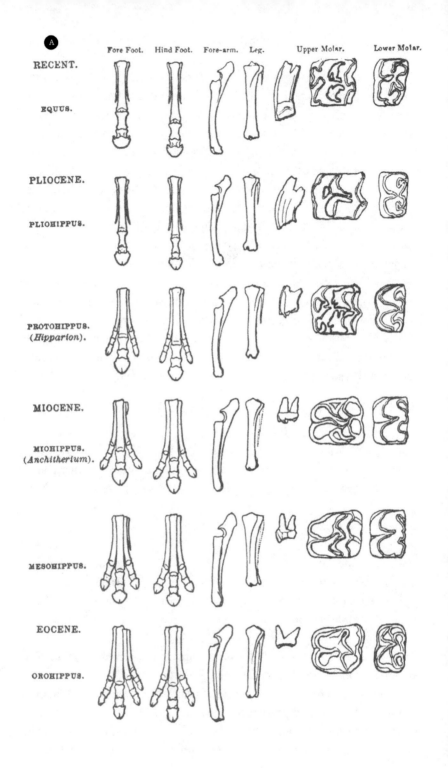

Figure 14.3 ◀ ▲

(A) Marsh's famous diagram of the changes in the limbs and teeth of North American fossil horses, and (B) Huxley's famous cartoon of "*Eohomo* riding *Eohippus*." (From Othniel Charles Marsh, "Polydactyl Horses Recent and Extinct." *American Journal of Science* 17 [1879]: 499–505.)

showed not only the changes in teeth but also in the skull, limbs, and even the toes. It became one of his favorite examples of an evolutionary series, and it has been used almost exclusively in nearly every book written on evolution or fossils since that time.

However, there were some problems with this evolutionary series. In the 1870s and even in the 1920s, the European sequence, *Hyracotherium-Anchitherium-Hipparion-Equus*, and the American sequence, *Eohippus-Orohippus-Mesohippus-Miohippus-Pliohippus-Equus*, were decent first approximations of horse evolution, but they were based on very incomplete samples. They fit the prevailing linear notion of evolution, with one lineage marching through time from *Eohippus* to *Equus*. As a simple way to convey the trends in the teeth, skull, limbs, and toes through time, this example is fine. But the dozens of additional species of fossil horses collected in the past century, which represent tens of thousands of fossils, have shown that the evolution of the horse is not a simple line from *Eohippus* to *Equus* but a complex bushy family tree, with many different lineages living at the same time (figure 14.4). This same pattern of a bushy, branching family tree is true of all known lineages that have a good fossil record: rhinos, camels, giraffes (chapter 15), elephants (chapter 16), and many others, even including humans (chapter 24). Nonetheless, the simplistic, outdated, linear view of horse evolution (see figure 14.1) still persists, even in textbooks and online media, no matter how much paleontologists try to correct it. A simple linear diagram certainly is easier to understand and remember, but the idea also fit the pre-Darwinian notion of the "chain of being" or "ladder of evolution." Old, outdated, but widespread concepts (like the name *Brontosaurus*, which was shown to be invalid in 1903) never seem to vanish from the popular mind, no matter how much new evidence comes to light and is widely publicized.

A lot has been learned since the 1920s about the earliest Eocene horses. When horses first emerged, they were so similar to their close relatives, the tapirs and the rhinos, that only a skilled eye could tell them apart (figure 14.5). I vividly remember my final project in an undergraduate vertebrate paleontology class at the University of California, Riverside, in 1975. My professor, Mike Woodburne, gave everyone in the class a random mixture of early Eocene mammal jaws and teeth from Emblem, Wyoming, in the Bighorn Basin, to simulate what we might have collected if we had taken a trip there. One of the hardest tasks for me was distinguishing the teeth of the earliest horses from their earliest tapir-rhino relative, *Homogalax*. It's just as hard to do if you have one of the rare skulls or jaws of these animals. Today any kid can tell a horse from a rhinoceros (and possibly recognize a tapir in a zoo), but 56 million years ago they would have looked nearly identical to our eyes. That's how much these three closely related groups have diverged through time as they adapted to different niches. The same could be said of the other early perissodactyl groups, such as the brontotheres. The last members of their lineage were elephantine beasts with huge paired bony blunt horns on their noses, but the earliest forms (like *Lambdotherium*) were dog-sized creatures that can barely be distinguished from a horse, a tapir, or a rhino. All of these early Eocene perissodactyls can now be traced to the Paleocene of Mongolia, where they seem to have a common ancestor in a

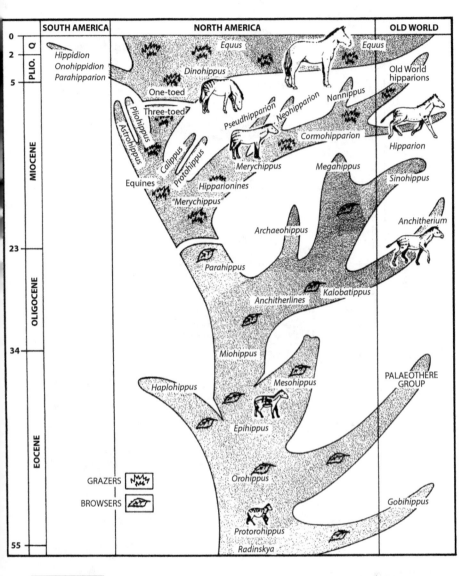

Figure 14.4 ▲

A modern view of horse evolution, emphasizing the bushy branching nature of their history, as many more fossils have been found and new species named. However, the overall trends toward higher-crowned teeth (shown by the symbols for browsing leaves or grazing grasses), larger size, longer limbs, and reduction of side toes are still true. (Drawing by C. R. Prothero)

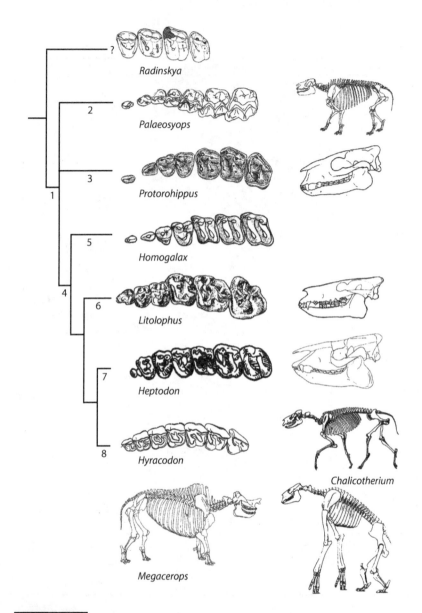

Figure 14.5 ▲

The evolutionary radiation of perissodactyls, showing the major branches of horses, rhinos, tapirs, chalicotheres, bronthotheres (*Megacerops*), and other extinct groups. As can be seen from the crown views of the upper left cheek teeth, the details of the crests and cusps are extremely similar between *Radinskya*, the early brontothere *Palaeosyops*, the primitive horse *Protorohippus* (long called *Hyracotherium*), the primitive moropomorph *Homogalax*, the chalicothere *Litolophus*, the tapiroid *Heptodon*, and the primitive rhinoceros *Hyracodon*. Shown next to the upper cheek teeth are typical skulls of horses, tapirs, and rhinos, emphasizing how similar they all looked in the early stages of perissodactyl evolution. The numbered branching points are as follows: (1) Perissodactyla, (2) Titanotheriomorpha, (3) Hippomorpha, (4) Moropomorpha, (5) Isectolophidae, (6) Chalicotherioidea, (7) Tapiroidea, (8) Rhinocerotoidea. (Redrawn by Carl Buell from several sources)

fossil *Radinskya* (named after the late Leonard Radinsky of the University of Chicago, who did a lot of pioneering work in early perissodactyls and tapirs). From there, perissodactyls and *Radinskya* appear be closely related to a group of archaic hoofed mammals called phenacodonts—and these can be traced back to the earliest hoofed mammals in the early Paleocene.

Going from the roots back up the family tree, we soon encounter the confused lineage surrounding the early Eocene horses *Hyracotherium* and *Eohippus*. These two animals were considered different and had different names until 1932, when British paleontologist Sir Clive Forster Cooper decided that they were the same animal. Because *Hyracotherium* was named first, its name was applied to all the early Eocene horses in North America, and most people learned it that way for decades. (A few American paleontologists still use that name today.) But in 1989, British Museum paleontologist Jerry Hooker restudied all the British *Hyracotherium* fossils accumulated since Owen's original report. He concluded that *Hyracotherium* was not a true horse at all but instead was related to a horse-like group called palaeotheres, which are well known from the early and middle Eocene of Europe. True horses only occurred in North America through most of their history until beasts like *Anchitherium* and *Hipparion* escaped to Eurasia in the Miocene. Then, in 2002, David Froehlich published a detailed analysis of all the species of early Eocene American horses and concluded that only one or two species could be referred to as part of Cope's old genus *Eohippus*, which had fallen out of use after 1932. The rest of the early Eocene horses belonged to a variety of genera, including old names like *Protorohippus* and *Pliolophus*, as well as some new genera, including *Sifrhippus*, *Arenahippus*, and *Minippus*. Referring to the earliest Eocene horses is complicated. You can't call them *Hyracotherium*, nor are most of them *Eohippus*; there are actually multiple genera. Most people hate complexity and prefer simple solutions, so they fall back on the old invalid names, either because they copied outdated diagrams or because they just don't know any better.

Throughout the rest of the Eocene, horses got slightly larger in size with creatures like *Orohippus* and *Epihippus*, followed by *Mesohippus* (see figure 14.4). *Mesohippus* fossils are particularly common in the upper Eocene and lower Oligocene beds of the Big Badlands of South Dakota and related deposits in Nebraska, Wyoming, Montana, Colorado, and North Dakota. Neil Shubin and I spent a lot of time looking at hundreds of these fossil horses in the American Museum of Natural History, where we were both students. We found that the old simplistic picture of linear evolution from *Mesohippus* to *Miohippus* was wrong. Instead, there were at least five different species of *Mesohippus* and four of *Miohippus*, and at one level in the rocks near Lusk, Wyoming, we found three species of *Mesohippus* and two species of *Miohippus* coexisting in the same place and time. Once again, horse evolution looked simple and linear when just a few fossils were known, but with large collections and detailed analysis, the species form a branching, bushy family tree, with multiple lineages overlapping in time. *Mesohippus* was about the size of German shepherd dog, and *Miohippus* was about the size of a Great Dane, with a longer muzzle than *Orohippus*, eyes slightly bigger and further back on the face, and a larger brain. In *Mesohippus*, the molar teeth were already more advanced, with well-developed cross-crests for chopping up leaves. Both horses had much longer, more slender legs than earlier ancestors, with only three toes on each hand and foot, and the side toes were much smaller than the weight-bearing middle toe. This middle finger becomes so broad that in *Miohippus* its upper end touches the wrist bone known as the cuboid, a landmark that distinguishes the two genera.

Multiple species of *Miohippus* lived throughout much of the Oligocene, until they were replaced by several genera including *Parahippus* and *Kalobatippus*. The early and middle Miocene saw an explosive radiation of horses (see figure 14.4). One lineage was the anchitherines, which retained the low-crowned cheek teeth of their ancestors, presumably because they lived in dense forests and subsisted on soft leaves and other browse. Even though their teeth remained primitive, the radiation of anchitherines such as *Anchitherium, Archaeohippus, Megahippus,* and *Hypohippus* evolved into a diverse range of body sizes. The largest of these, *Megahippus* and *Hypohippus*, even reached the size of a modern horse, but their huge teeth still retained the low-crowned cusps of the earliest horses.

But the main radiation of horses went in another direction, developing cheek teeth with deep roots so they would never run out of grinding surface as they wore their teeth down. Grass contains gritty pieces of silica called phytoliths that quickly wear teeth down, and it had been assumed that this heavy tooth wear was due to a diet of grasses. In recent years, that story has been modified because huge areas of modern grassland did not appear until much later in the Miocene. However, another source of tooth wear is gritty sand and silt that sticks to the grass stalks, and that could explain why horses (and camels and rhinos and many other Miocene groups) evolved high-crowned teeth before the huge expansion of modern grasslands.

Multiple lineages of horses overlapped at the same time in the Miocene: the radiation of different horses that used to be lumped into the wastebasket genus *Merychippus* (now split into many genera); the huge radiation of the hipparionine horses, which spread to Eurasia and became very successful; plus dwarf horses like *Calippus* and *Nannippus*; and many more. In Railway Quarry A in the Valentine Formation of north-central Nebraska, 12 different species of horse were pulled from the same hole in the ground, indicating that many species were living in the same place and in the same time (one anchitherine, several hipparionines, and many others). Most of these horses were still three-toed, although the side toes were tiny and normally did not touch the ground, whereas the middle toe was robust and carried most of the weight. Their limbs were also much longer and more slender, and their skulls look much like that of a modern horse, although they were not as big.

Finally, in the late Miocene, nearly all three-toed horse lineages vanished, and only the one-toed lineage from *Dinohippus* (not *Pliohippus*, as in the old diagrams) to *Equus* remained in the Pliocene. During the Ice Ages, the genus *Equus* underwent yet another huge radiation, but until recently the Ice Age horses (especially in North America) were split into dozens of different species, mostly based on small differences in the teeth and the proportions of the skulls and limbs. The classification of Ice Age American horses was in chaos, with some people recognizing lots of species and others lumping the variation into just a few species.

Beginning in 2005, and in subsequent years, molecular biology jumped into the fray. The excellent preservation of some of the youngest Ice Age horse fossils enabled recovery of both nuclear DNA and mitochondrial DNA from a number of extinct species. For many living species, the answer confirmed what had already been suggested by their anatomy and behavior. The asses and onagers and kiangs of Eurasia and Africa were all closely related, as were the four species of zebra (one of which, the quagga, was driven to extinction in South Africa just over a century ago). These results showed that we could rely on the molecular data because it exactly matched the evolutionary history that had already been worked out based on fossils and living animals.

But for the North American horses, the answer stunned and surprised everyone. The genetic evidence supported only a few species of North

American Ice Age horses, not the dozens that already had been named. In addition, they were all close relatives of the lineage of the living horse, *Equus caballus*. Genetic evidence also confirmed that modern horses were descendants of Przewalski's horse, which still roams the steppes of Asia and Siberia. The one exception to this discovery concerned the extinct stilt-legged horses, which have very long slender limbs compared to caballine horses, zebras, and asses. Even though they were separated into several different species based on fossils, the genetic evidence showed they were a distinct genus and species that split from the main horse lineage about 5 million years ago. Scientists renamed this horse *Haringtonohippus*, after Canadian paleontologist Richard Harington who first recognized its uniqueness. Today there are about nine species of living wild horses in the genus *Equus*, including three species of zebra (plus the quagga), three or four species of asses and onagers, plus Przewalski's horse.

During the Ice Ages, horses spread not only from their North American homeland to Eurasia to form the great radiation of living horses and asses but also to Africa, where they evolved into zebras. They even spread to South America, where a bizarre group of horses, the hippidions, apparently evolved a short proboscis like that of a tapir. At the end of the last Ice Age, about 10,000 years ago, horses vanished from the Americas, along with most other large mammals, including mammoths, mastodonts, ground sloths, saber-toothed cats, and many others. The reasons were complex and probably involved both climate change and possibly some human hunting. Horses did not return to their ancestral North American homeland until 1493 when Columbus brought them from Spain on his second voyage. Today wild mustangs and all the domestic breeds of horses are found in the Americas, and Plains Indians relied on horses as the basis for their nomadic culture— but these are recent developments.

But did the side toes of horses really vanish? Every once in a while a horse is born with side toes that are somewhat like those of the Miocene horses (figure 14.6A). They are not a tremendous burden on the horse and don't seem to affect their running speed, but they do look odd. They have been nicknamed "horned horses," and their strangeness has often garnered attention. Julius Caesar is said to have ridden on a horned horse as a way of looking more regal and almost mythological. This kind of development, or evolutionary throwback, is known as an *atavism*. All living horses still have the genes for side toes, but normally their gene regulation shuts off

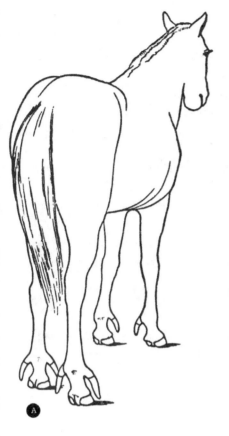

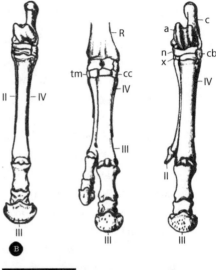

Figure 14.6 ▲

(A) A famous example of a rare mutant horse that has three toes, rather than one. (B) Bone structure of the feet of the mutant horses. On the left is a normal horse foot; in the middle is an extra toe formed by duplicating the central toe; and on the right is an extra toe formed by enlarging the reduced side toes (splint bones), which were functioning side toes in earlier horses. (From Othniel Charles Marsh, "Recent Polydactyle Horses," *American Journal of Science* 43 [April 1892]: 340–355)

the instructions for making side toes. We never see them expressed unless the regulation fails, and then we get a three-toed horse. Modern horses also have tiny splints of their side toes running alongside their robust central toe (figure 14.6B). These splints serve no real function and are useless leftovers not yet quite removed by evolution. If a horse breaks one of these side toes, it can be crippled for life, which is a strong selective pressure against the genes for tiny side toes.

The story of the evolution of the horse is far more complex than how it was portrayed in the 1870s or even in the 1920s. But it is still an outstanding example of how the fossil record can reveal the detailed ancestry of a living animal.

FOR FURTHER READING

Franzen, Jens Lorenz. *The Rise of Horses: 55 Million Years of Evolution*. Baltimore, Md.: Johns Hopkins University Press, 2010.

MacFadden, Bruce J. *Fossil Horses: Systematics, Paleobiology, and the Evolution of the Family Equidae*. Cambridge: Cambridge University Press, 1992.

Prothero, Donald R., *The Princeton Field Guide to Prehistoric Mammals*. Princeton, NJ: Princeton University Press, 2016.

Prothero, Donald R., and Robert M. Schoch, eds. *The Evolution of Perissodactyls*. Oxford: Oxford University Press, 1989.

——, eds. *Horns, Tusks, and Flippers: The Evolution of Hoofed Mammals*. Baltimore, Md.: Johns Hopkins University Press, 2002.

Rose, Kenneth D., and J. David Archibald, eds. *The Rise of Placental Mammals: Origin and Relationships of the Major Extant Clades*. Baltimore, Md.: Johns Hopkins University Press, 2002.

Savage, Robert J. G., and M. R. Long. *Mammal Evolution: An Illustrated Guide*. New York: Facts-on-File, 1986.

Turner, Alan, and Mauricio Anton. *National Geographic Prehistoric Mammals*. Washington, D.C.: National Geographic Society, 2004.

HOW THE GIRAFFE GOT ITS NECK

The giraffe, by its lofty stature, much elongated neck, fore-legs, head and tongue, has its whole frame beautifully adapted for browsing on the higher branches of trees. It can thus obtain food beyond the reach of the other Ungulata or hoofed animals inhabiting the same country; and this must be a great advantage to it during dearths. . . . . So under nature with the nascent giraffe the individuals which were the highest browsers, and were able during dearth to reach even an inch or two above the others, will often have been preserved; for they will have roamed over the whole country in search of food. . . . Those individuals which had some one part or several parts of their bodies rather more elongated than usual, would generally have survived. These will have intercrossed and left offspring, either inheriting the same bodily peculiarities, or with a tendency to vary again in the same manner; whilst the individuals, less favoured in the same respects will have been the most liable to perish. . . . By this process long-continued, which exactly corresponds with what I have called unconscious selection by man, combined no doubt in a most important manner with the inherited effects of the increased use of parts, it seems to me almost certain that an ordinary hoofed quadruped might be converted into a giraffe.

—CHARLES DARWIN, *ON THE ORIGIN OF SPECIES* (1859)

One of the most amazing of all animals is the giraffe. We all grow up with pictures of them in our children's books, and most of us have seen them in the zoo. People have been amused and mystified by them ever since they first saw them. The San people of southern Africa have a medicine dance called the "giraffe dance" that supposedly treats head ailments. In African folk tales, the giraffes grew tall from eating magical herbs. Giraffes appear in ancient African art of the Egyptians, Kushites, and Kiffians. The Kiffians are thought to have been the people that produced a life-sized rock engraving of two giraffes that is considered the world's largest petroglyph. The Egyptians even had a hieroglyph for the giraffe, kept giraffes as pets, and shipped them around the Mediterranean to other cultures.

When the Romans first encountered them, they thought giraffes were hybrids of a long-necked camel and a spotted leopard and called them "camelopards"—a name that lives on in their scientific name, *Giraffa camelopardalis*. In 46 BCE, Julius Caesar obtained one and displayed it in Rome, the first time Romans had ever seen this bizarre creature. Numerous other giraffes were brought to Rome, often to take part in parades or to be slaughtered in the arena with other humans and animals. Once the Roman Empire fell, giraffes vanished from the European consciousness and had a mythic status in the Middle Ages, although some Europeans knew of them through contact with Arab traders. In 1414, Arab merchants shipped a giraffe from Bengal to China, where it came to live in the emperor's menagerie. It fascinated the Chinese, who thought it was related to a mythical horned hoofed creature called the Qilin. In 1486, a giraffe was presented to Lorenzo de Medici, the ruler of Florence during its greatest heyday in the Renaissance. In the early 1700s, Prince Muhammad Ali of Egypt gave a specimen named "Zarafa" to the French king Charles X as a goodwill token. It caused a sensation in Paris, and the mania for all things giraffe came to be known as "giraffanalia." In modern times, giraffes have appeared in Salvador Dalí paintings, in numerous children's books and movies, and as lots of famous characters, including Melman the neurotic giraffe (voiced by David Schwimmer) in the *Madagascar* movies and even Geoffrey the Giraffe, mascot for the now bankrupt Toys "R" Us stores.

The image of the long-necked giraffe has long been a staple of Western culture, and so have been attempts to explain it. In 1809, Jean-Baptiste de Monet, the Chevalier de Lamarck, used the neck of the giraffe (figure 15.1) as one of his examples of how creatures change through time:

> It is interesting to observe the result of habit in the peculiar shape and size of the giraffe: this animal, the tallest of the mammals, is known to live in the interior of Africa in places where the soil is nearly always arid and barren, so that it is obliged to browse on the leaves of trees and to make constant efforts to reach them. From this habit long maintained in all its race, it has

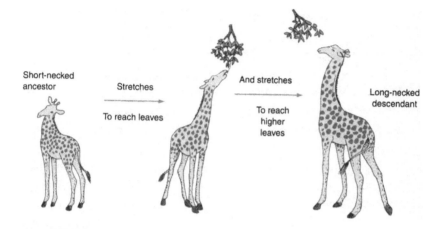

Figure 15.1 ▲

Diagram showing Lamarck's conception of how the giraffe obtained its long neck from stretching and stretching for higher leaves in each generation, and then passing those variations on to the next generation.

resulted that the animal's forelegs have become longer than its hind-legs, and that its neck is lengthened to such a degree that the giraffe, without standing up on its hind-legs, attains a height of six meters.

This short passage from Lamarck is often used to epitomize his entire complex theory of how life evolves, and it is constantly cited in biology textbooks as well. Like most naturalists of his time, Lamarck thought changes made in our adult bodies could be passed on directly to our descendants (known as the "inheritance of acquired characters"). According to his commonly held idea, the strong muscles of a blacksmith would be passed on to his sons, and the neck stretching for the highest leaves in a tree by ancestral short-necked giraffes would be passed on to their descendants. In Lamarck's words: "variations in the environment induce changes in the needs, habits and modes of life of living beings . . . these changes give rise to modifications or developments in their organs and the shape of their parts."

Sadly, this is often the only part of Lamarck's complex notions about evolution still remembered today, largely because Lamarck's rival, Baron Georges Cuvier, did his best to destroy and distort Lamarck's legacy after the great man died. In fact, before the discovery of the mechanism of Mendelian inheritance and early genetics in the late 1800s and early

1900s, most naturalists believed in the inheritance of acquired characters—including Charles Darwin.

Today Lamarck is mocked as the "guy who got evolution wrong" even though so many of his ideas were groundbreaking and on the right track. In addition, it was Lamarck who recognized the unity of zoology and botany and even coined the term "biology." Lamarck survived the French Revolution and the Reign of Terror, but he was forced to switch from botany and take the undesirable post of curator of "Insects and Worms" at the Natural History Museum in Paris. In the process, he ended up revolutionizing and essentially creating our modern field of invertebrate zoology. Unfortunately, the idea of "Lamarckism" or "Lamarckian inheritance" has been reduced to just one wrong idea that was a minor part of his thinking, an idea that was held by every naturalist for many more decades.

Of course, Charles Darwin tried to explain the neck of the giraffe in his book on evolution as well (see chapter beginning epigraph). However, Darwin's idea was quite different from Lamarck's. According to Darwin, the most important idea is that natural populations are highly variable, and no two individuals are truly alike, even if they are siblings (except for identical twins). For example, among the ancestral population of giraffes might be some with slightly longer necks. During periods of drought when vegetation is scarce, they would be able to reach leaves higher on the trees than other giraffes in the population and thus survive. These longer-necked ancestral giraffes then become the parents of the next generation, passing on the genes for longer necks to their descendants. After many generations of doing this, the population of giraffes would have necks that on average were longer and longer (figure 15.2).

This story is very appealing and seems self-evident, but research on living giraffe behavior shows that it is seldom true. In fact, most of the time giraffes feed with their neck held horizontally, and they also feed a lot on ground vegetation with their head down and reaching to their feet. Only rarely are giraffes seen reaching for the highest parts of the trees because most parts of the trees are too high for any other browser to reach. In a 2010 study, it was found that the longest necked individuals died more often than shorter individuals in drought conditions because a longer neck required more nutrients, which are scarce during a drought.

Instead, giraffe biologists have found that longer necks mostly convey reproductive advantage for the males, who tend to be significantly taller

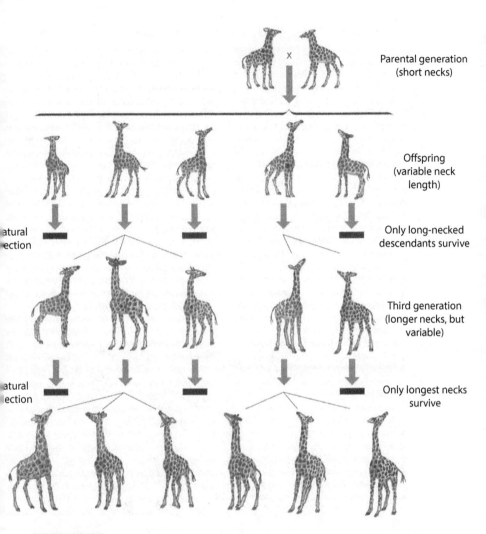

Parental generation
(short necks)

Offspring
(variable neck
length)

Natural
selection

Only long-necked
descendants survive

Third generation
(longer necks, but
variable)

Natural
selection

Only longest necks
survive

Figure 15.2 ▲

Darwin suggested that natural selection explains the long neck of the giraffe. Each generation has a variety of neck lengths, but during hard times, only the longer-necked individuals can reach food and survive, and they pass those genes for longer necks on to the next generation until all the giraffe populations have longer necks.

than the females. They compete for mates with other rival males, often engaging in violent battles that involve a lot of pushing and shoving and slamming their neck against their rivals ("necking"), and sometimes injuring them. Overall size and a long neck does give a bigger male giraffe a mating advantage over rival smaller males.

Indeed, having such a long neck is very complicated and difficult to maintain. To pump blood up such a long neck to the head requires a huge heart that weighs more than 11 kilograms (25 pounds). It creates almost double the blood pressure that a human heart can produce. The giraffe heart must contract at more than 150 beats per minute to maintain this pressure, much higher than the normal resting human pulse rate of about 90. It's hard enough to pump all that blood uphill to the brain, but it is an even bigger challenge when they lower their head (figure 15.3). Like humans, the giraffe would faint if all the blood rushed to its head at once. So giraffes have a network of fine blood vessels, the *rete mirabile* (wondrous net), that reduces the pressure of

Figure 15.3 ▲

Giraffes have a challenge reaching down to the ground to drink or feed. (Courtesy of Wikimedia Commons)

the blood to the brain. In addition, a giraffe might faint from excess flow of blood were it not for seven one-way valves in the jugular vein. This prevents the blood returning to the heart from the brain from pouring back to the head when it is lowered. All this high blood pressure in the neck and forelimbs forces the giraffe to have very thick, tight skin in its neck and front legs to hold in the pressure, just like compression hose are worn by people with circulatory problems. (Scientists have studied giraffe neck skin to determine its properties and see if it can be duplicated artificially for industrial purposes.)

All of these features are essential for their survival as long-necked animals, so some might say that the giraffe is well designed. But that does not explain the course of the left recurrent laryngeal nerve, which in humans connects the brain to the larynx and allows us to speak. In all mammals, this nerve avoids the direct route between brain and throat and instead descends into the chest, loops around the aorta near the heart, then returns to the larynx (figure 15.4). That makes it 7 times longer than it needs to be! For an animal like the giraffe, it traverses the entire neck twice, so it is 15 feet long (14 feet of which are unnecessary!). Not only is this design wasteful, but it also makes an animal more susceptible to injury.

The bizarre pathway of this nerve makes perfect sense in evolutionary terms, however. In fish and early mammal embryos, the precursor of the left recurrent laryngeal nerve attached to the sixth gill arch, deep in the neck and body region. Fish still retain this pattern, but during later mammalian embryology the gill arches were modified into the tissues of our throat region and pharynx. Parts of the old fish-like circulatory system were rearranged, so the aorta (also part of the sixth gill arch) moved back into the chest, taking the left recurrent laryngeal nerve (looped around it) backward as well. Giraffes inherited this clumsy, poorly designed system because it was part of their embryonic and evolutionary past. There was no great cost to keeping it that way, and it was almost impossible to rearrange it due to anatomical constraints. For the giraffe, the nerves fire in microseconds, so the difference in their reaction time of a nerve impulse going 15 feet rather than 1 foot is negligible. This is a classic example of how inefficient and poorly designed nature can be when it blindly follows embryonic pathways established by our fish-like ancestors. Despite this inefficiency, the giraffe gets along fine with an extra 14 feet of nerves. As long as giraffes can survive and breed to leave more giraffes, the inefficiency of their neck nerves matters less than the long neck that gives them a survival advantage.

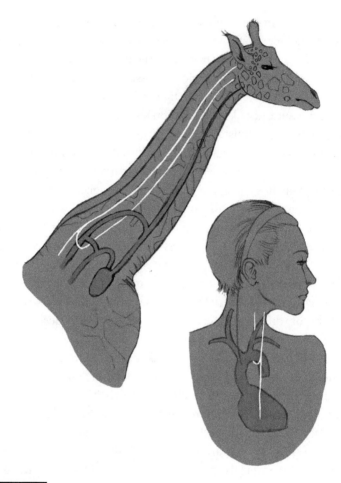

Figure 15.4 ▲

The left recurrent laryngeal nerve in a human runs from the spinal column, down past the aorta on the heart, then up to the voice box. The course in the giraffe is the same, making it 15 feet long just to travel what would only be a few inches in a direct line. (Drawing by Mary Persis Williams)

From Lamarck to Darwin, all of the ideas about how giraffes got their long necks were pure speculation until recently because no one had found fossils of giraffes. But early in the twentieth century, numerous giraffe fossils were found in Miocene (23–5 million years old) beds all over Africa and southern Asia. Today there are at least 24 genera and many different species of extinct giraffids. The big surprise is that *all of them* had short necks! In fact, the Miocene was the time of an enormous radiation of short-necked giraffes

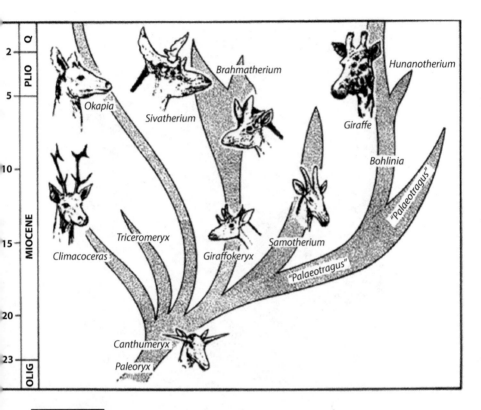

Figure 15.5 ▲

Evolution of the giraffe family. The modern okapi is more typical of the group, with its short neck and relatively short horns or "ossicones." Some fossil giraffids, however, had very unusual branching and flaring cranial appendages. Only the lineage of the modern giraffid evolved a long neck. (Drawing by C. R. Prothero)

(figure 15.5), often with bizarre deer-like or moose-like or antelope-like horns (living giraffes have short cylindrical horns called ossicones). Some of them, like *Prolibytherium*, had moose-like horns in the males. Others had horns like deer or antelopes. The huge *Sivatherium* was more than 10 feet tall, with thick forked horns, and weighed half a ton. The giant *Brahmatherium* was almost as tall as a modern giraffe but had a short thick neck and enormous horns on its head. All of them have relatively short necks.

In retrospect, this should not surprise us because the other genus of living giraffid is the okapi, a reclusive forest browser from the Congo jungles with a short neck (figure 15.6). First discovered and named as recently as 1901, it was one of the last large mammals to be discovered, and it was

Figure 15.6 ▲

The only other living giraffid is the okapi, which lives in the dense jungles of the Congo basin. (Courtesy of Wikimedia Commons)

rumored to be the "African unicorn." The okapi is sometimes called the zebra giraffe because the rear haunches and legs are striped like a zebra, providing camouflage in the dense jungles. Living a mostly solitary life browsing on dense jungle leaves, it is silent and seldom seen and today is considered endangered because of excessive poaching and destruction of its forest habitat. The pressure from poachers is vicious because there is a lucrative black market for okapi parts. In June 2012, a gang of poachers attacked the headquarters of the Okapi Wildlife Reserve in the Demo-cratic Republic of the Congo, killing six guards and other staff, as well as all 14 okapis that were in their captive breeding center.

We now have a fossil record that shows how giraffids evolved and demonstrates that they are far more diverse than the single long-necked genus *Giraffa* we all know. Still some people did not expect that we could find fossil evidence of how *Giraffa* got its long neck. In 2009, my friend Nikos Solounias and his students described fossils of an extinct Asian giraffid called *Samotherium major*. Unlike the incomplete fossils of most extinct giraffids, this one had a complete neck—and it was halfway in length between the neck of the okapi and the modern *Giraffa* (figure 15.7). One could not ask for a more perfect intermediate form!

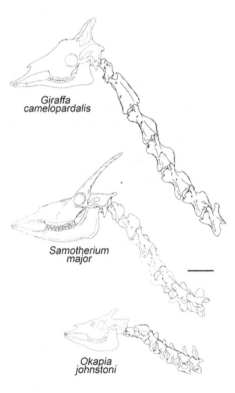

*Giraffa
camelopardalis*

*Samotherium
major*

*Okapia
johnstoni*

Figure 15.7 ▲

The recently discovered neck and head of the giraffid with an intermediate length neck, *Samotherium major* (*center*), compared to the okapi (*bottom*) and the long-necked *Giraffa* (*top*). (Courtesy of Nikos Solounias)

The story of the giraffe is interesting and complex. We know their long necks evolved, and we now have the fossils to show how it happened. In fact, we have fossils of more than two dozen extinct types of giraffids, almost all of which had short necks and weird headgear. The lengthening of the giraffe's neck was not due entirely to the processes postulated by Lamarck and by Darwin, despite decades of textbook orthodoxy. And the course of the left recurrent laryngeal nerve in their long neck provides a great example of a clumsy and inefficient design, which belies attempts to point to a perfect designer.

FOR FURTHER READING

Danowitz, Melinda, Aleksandr Vasilyev, Victoria Kortlandt, and Nikos Solounias. "Fossil Evidence and Stages of Elongation of the *Giraffa camelopardalis* Neck." *Royal Society Open Science* 2, no. 10 (2015): 150393. http://doi.org/10.1098/rsos .150393.

Mitchell, G., and J. D. Skinner. "On the Origin, Evolution, and Phylogeny of Giraffes, *Giraffa camelopardalis.*" *Transactions of the Royal Society of South Africa* 58, no. 1 (2003): 51–73.

Prothero, Donald R., *The Princeton Field Guide to Prehistoric Mammals.* Princeton, NJ: Princeton University Press, 2016.

Prothero, Donald R., and Robert M. Schoch. *Horns, Tusks, and Flippers: The Evolution of Hoofed Mammals.* Baltimore, Md.: Johns Hopkins University Press, 2003.

HOW THE ELEPHANT GOT ITS TRUNK

Then the Elephant's Child sat back on his little haunches, and pulled, and pulled, and pulled, and his nose began to stretch. And the Crocodile floundered into the water, making it all creamy with great sweeps of his tail, and he pulled, and pulled, and pulled. And the Elephant's Child's nose kept on stretching; and the Elephant's Child spread all his little four legs and pulled, and pulled, and pulled, and his nose kept on stretching; and the Crocodile threshed his tail like an oar, and he pulled, and pulled, and pulled, and at each pull the Elephant's Child's nose grew longer and longer—and it hurt him hijjus! Then the Elephant's Child felt his legs slipping, and he said through his nose, which was now nearly five feet long, 'This is too butch for be!' Then the Bi-Coloured-Python-Rock-Snake came down from the bank, and knotted himself in a double-clove-hitch round the Elephant's Child's hind legs, and said, 'Rash and inexperienced traveller, we will now seriously devote ourselves to a little high tension, because if we do not, it is my impression that yonder self-propelling man-of-war with the armour-plated upper deck' (and by this, O Best Beloved, he meant the Crocodile), 'will permanently vitiate your future career.' That is the way all Bi-Coloured-Python-Rock-Snakes always talk. So he pulled, and the Elephant's Child pulled, and the Crocodile pulled; but the Elephant's Child and the Bi-Coloured-Python-Rock-Snake pulled hardest; and at last the Crocodile let go of the Elephant's Child's nose with a plop that you could hear all up and down the Limpopo.

—RUDYARD KIPLING, "THE ELEPHANT'S CHILD" (1902)

Elephants are truly amazing creatures, with their enormous size, incredible intelligence, giant ears, and a trunk that serves as a multifunctional tool for them. Asian elephants have been important work animals for millennia, and they were used in warfare in ancient times to break up enemy infantry formations. They have been important cultural symbols for a long time as well, especially in the civilizations of Africa and southern Asia, which were in close contact with wild elephants. The Hindu god Ganesha had the head of an elephant, and they are widely depicted in the art of many cultures. Elephants are often thought to symbolize strength, power, wisdom, longevity, stamina, leadership, sociability, nurturance, and loyalty. In popular Western culture, they are most familiar to us from circuses and zoos, and they even became the symbol of the Republican Party. They are ubiquitous in literature, from Kipling's *Just-So Stories* (see chapter epigraph), the Babar stories, Disney's Dumbo, and Dr. Seuss's Horton.

The largest living land mammals, few natural predators can harm elephants except when they are young. In the wild, they can live 60 to 70 years, but they seldom live to full adulthood in most places due to poaching. As late as 1979, there were as many as 3 million elephants in Africa alone, yet their populations are now down 90 percent or more in most of Africa due to relentless poaching for their ivory, which is more valuable pound for pound than cocaine. Despite a worldwide effort to ban the sale of ivory, the black market remains very lucrative, and political instability in Africa (along with abundant weapons from numerous wars) makes it difficult to protect elephants from poachers except in a few well-guarded reserves. As long as there is a huge demand for ivory in newly prosperous China and Vietnam, there is little hope that elephants will survive much longer in the wild outside of those in a few reserves.

As much as elephants have become familiar to us, where they came from, who they were related to, and how they got their amazing trunks had long been a mystery. There were lots of myths and legends among the cultures that knew them well in Asia and Africa, but there was little science to back it up. Rudyard Kipling's famous collection of bedtime tales, *Just-So Stories*, included the famous account of how "The Elephant Child" got its trunk when a crocodile grabbed its short floppy nose and stretched it out (figure 16.1). Although fossil elephants and their relatives were among the earliest finds in the history of paleontology, full evidence for the earliest elephant relatives and elephants' origins only became available recently.

Fossils of elephants were known to the ancients, and their huge skulls with a central nasal opening were mistaken for the eye openings of the giant one-eyed "Cyclops." Their huge bones were often collected when they were found, and the Roman emperor Augustus had a collection of bones that

Figure 16.1 ▲

An illustration from Rudyard Kipling's "The Elephant's Child." (Courtesy of Wikimedia Commons)

were found near his villa in Capri (across the bay from Pompeii and Naples). The Roman historian Pliny recorded huge pieces of ivory found in the ground, and during the Middle Ages the huge bones were attributed to giant biblical humans in Genesis 6:4: "There were giants in the earth in those days."

By the late 1700s, however, it became harder and harder to explain these huge bones as the remains of giant biblical humans. In 1739, Charles le Moyne, the second Baron de Lougueil, left Montreal with French and Indian troops to fight Chickasaw Indians along the Ohio River. Somewhere along the way he found what appeared to be the remains of three elephants. After the war ended a year later, he went back to collect them and sent them to New Orleans. Ultimately they reached Paris and came to the attention of French naturalists. In the 1740s and 1750s, English settlers sent more remains from this locality (known today as Big Bone Lick, Kentucky), and they were seen by Benjamin Franklin in America and reached England as well. The large limb bones and tusks clearly resembled elephants, but the odd teeth were a mystery (figure 16.2). They were clearly unlike any living elephant, yet they were part of an animal of elephantine size. (We now know that these are teeth of the American mastodon.) Franklin speculated that the teeth were reminiscent of a carnivorous animal, but later he and others decided it was

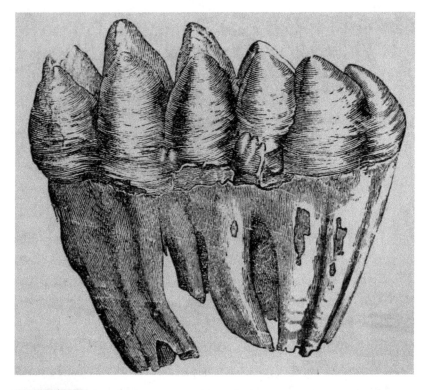

Figure 16.2 ▲
The controversial, mysterious mastodon molar tooth known as the "Great Incognitum."
(Courtesy of Wikimedia Commons)

a vegetarian. In 1769, the famous British anatomist William Hunter took Franklin's idea seriously. He suggested that it was not a true elephant but a "pseudelephant," or American *incognitum* (Latin for "unknown"), that had independently developed ivory tusks: "This monster, with the agility and ferocity of a tiger . . . cruel as the bloody panther, swift as the descending eagle, terrible as the angel of right . . . and if this animal was indeed carnivorous, which I believe cannot be doubted, we may as philosophers regret it, as men we cannot but thank Heaven his whole generation is probably extinct."

The idea that any animal had ever become extinct was anathema to the learned men of that time because it violated the idea of Divine Providence. As Alexander Pope wrote in his famous *Essay on Man,* "Who sees with equal eye, as God of all, a hero perish or a sparrow fall." Extinction of any creature would break the continuous "Great Chain of Being" that God had ordained.

In Pope's words, "Where, one step broken, the great scale's destroy'd; from Nature's chain whatever link you strike, Ten or Ten thousandth, breaks the chain alike." Yet more and more fossils of huge elephant-like creatures that were clearly no longer around in Europe kept turning up, not only the American *incognitum* but also frozen carcasses of Siberian mammoths, as well as abundant teeth and ivory and bones of these creatures. George Louis Leclerc, the Comte de Buffon, concluded in his *Théorie de la Terre* in 1749 that most of the supposedly extinct animals were hiding somewhere in an unknown region but it was likely that the large terrestrial mammals such as the mammoth and the *incognitum* had perished. By 1778, Buffon was relating their disappearance to his ideas of violent cataclysms in Earth's early history. During this time, the climate was warmer, the polar regions were tropical, and elephants (meaning mammoths) lived in Siberia. This implied an Earth of much greater antiquity than biblical accounts described. Buffon suggested that Earth was as much as 75,000 to 3 million years old, rather than the 6,000 years demanded by most literalist biblical scholars. Naturally, such revolutionary ideas were not popular with the theologians in the Sorbonne. Buffon was protected by the king, however, so he was not persecuted for his heresy, although his ideas were not widely accepted either.

U.S. president Thomas Jefferson was not only a politician and a writer and a leader but also an avid naturalist and fossil collector. A believer in the Great Chain of Being, in 1799 he wrote:

> The bones exist: therefore the animal has existed. The movements of nature are in a never ending circle. The animal species which has once been put into a train or motion, is still probably moving in that train. For if one link in nature's chain might be lost, another and another might be lost, till this whole system of things should vanish by piece-meal; a conclusion not warranted by the local disappearance of one or two species of animals, and opposed by the thousands and thousands of instances of the renovating power constantly exercised by nature for the reproduction of all her subjects.

He was sure that the great *Incognitum* must still live in the unexplored wilds of western North America. As he wrote in 1781:

> It may be asked, why I insert the Mammoth, as if it still existed? I ask in return, why I should omit it, as if it did not exist? Such is the economy of nature, that no instance can be produced of her having permitted any one race of her animals

to become extinct; of her having formed any link in her great work so weak as to be broken. To add to this, the traditionary testimony of the Indians, that this animal still exists in the northern and western parts of America, would be adding the light of a taper to that of the meridian sun. Those parts still remain in their aboriginal state, unexplored and undisturbed by us, or by others for us. He may as well exist there now, as he did formerly where we find his bones. . . . It would be erring therefore against that rule of philosophy, which teaches us to ascribe like effects to like causes, should we impute this diminution of siace [of animals] in America to any imbecility or want of uniformity in the operations of nature. . . . Animals transplanted into unfriendly climates, either change their nature and acquire new fences against the new difficulties in which they are placed, or they multiply poorly and become extinct.

When delays prevented William Lewis and Meriwether Clark from leaving until 1803, President Jefferson instructed them to go to Big Bone Lick and collect some more of the mysterious beast. When he received gigantic claws from some cave deposits, he instructed Lewis and Clark to look for a gigantic lion during their expedition to the great Northwest. (The claws turned out not to be of a lion but of a giant ground sloth, now named *Megalonyx jeffersoni*.)

In 1779, the German naturalist Peter Simon Pallas, working in St. Petersburg, Russia, described the frozen carcass of a rhinoceros in Siberia. He concluded that this was "convincing proof that it must have been a most violent and most rapid flood which once carried these carcasses toward our glacial climates, before corruption had time to destroy their soft parts." (We now realize that this was the woolly rhinoceros, a cold-adapted species.)

The fact of extinction was finally proven by one of the greatest biologists of all time, Baron Georges Cuvier. He was an outstanding figure in French science, and he survived the reign of Louis XVI, the French Revolution and Reign of Terror, Napoleon, and subsequent French kings without loss of status or position. Cuvier became the founder of comparative anatomy and of vertebrate paleontology, developing tremendous skill in describing and recognizing the bones of vertebrates. He is most famous for his "law of correlation of parts." This was simply the observation that vertebrate anatomy has many predictable patterns, depending on habitat and diet of the animal. For example, a predator not only has sharp meat-cutting teeth but also sharp claws, whereas a herbivorous mammal usually has grinding teeth and hooves. An apocryphal story claims that one night a prankster burst into his bedchamber dressed as the Devil and told him that he would be eaten alive. Cuvier allegedly replied, "You cannot eat me. You have horns and hooves, so you must eat plants."

In 1796 Cuvier read a paper before the French Institute on living and fossil elephants. He was the first scientist to recognize the difference between the Asian and African elephants. Then he showed that the mammoth and the American *incognitum*, although related to elephants, were not the same as living elephants and did not require the earth to be much warmer in the past. He pointed to these fossils and other large animals—the Siberian rhinoceros mummy, the giant ground sloth, and the first mosasaur fossil— and said they were too huge not to be found yet in a world that was rapidly becoming explored. They "prove the existence of the world before ours, and [were] destroyed by some catastrophe." Cuvier went on to develop his ideas of a great catastrophe that had preceded our world but was not mentioned in Genesis. This was the "antediluvian" world before Noah's flood, a time of darkness, great monsters, and cataclysmic changes. The fact of extinction was proven beyond a doubt, although Cuvier's notion of the antediluvian world soon crumbled as details of the fossil record became better known.

Throughout the rest of the 1800s, more and more mammoth and mastodon skeletons were found, establishing that these great beasts roamed all of the northern continents during the Ice Ages (a concept that first emerged in 1837). The Miocene rocks of Europe also yielded more primitive proboscideans, or elephant relatives. One of these was the huge *Deinotherium* ("terrible beast" in Greek), named by Johann Jacob von Kaup in 1831. It was a proboscidean bigger than the biggest mammoth, but with only two tusks, both in its lower jaw, that curved downward. Other odd mastodonts were found in Eurasia and North America, showing that the proboscideans had roamed the Northern Hemisphere ever since the early Miocene, about 18 million years ago. But no proboscidean fossils older than these were found in Eurasia or North America. Where did elephants come from?

The solution came with one of the first paleontological expeditions to Africa. In 1902, British paleontologists C. W. Andrews and H. J. L. Beadnell led an expedition to explore the Fayûm Basin of Egypt just west of the Great Pyramids. In addition to fossils of many weird beasts that lived only in Africa and are now extinct, they found fossils not only of some of the oldest monkey-like primates but also of early relatives of elephants. They found numerous teeth of a hippo-sized beast called *Barytherium* ("heavy beast" in Greek), which turned out to be a slightly more advanced proboscidean with fully developed cross-crests on its molars, a high forehead, and a short trunk. But the most spectacular find was a complete skeleton of a late Eocene fossil called *Moeritherium* ("beast of Lake Moeris," a dry lake near Fayûm). It was the size and shape of a pygmy hippopotamus or tapir, about 0.7 meters (2.3 feet) tall but almost 3 meters (10 feet) long. *Moeritherium* had a barrel-shaped body and short stout legs. It had short tusks in its upper and lower canines, and a short proboscis like that of a tapir as well (figure 16.3).

The main lineage (figure 16.4) of proboscideans was represented by early Oligocene fossils, again from the Fayûm beds of Egypt but later than late Eocene *Moeritherium*. The best known of these are *Palaeomastodon* and *Phiomia*. About 2 meters (6 feet 5 inches) high at the shoulder and weighing as much as a small rhinoceros, *Palaeomastodon* had a long jaw

Figure 16.3 ▲

One of the earliest proboscideans was *Moeritherium*, which had a pig-like body with a tapir-like snout. (Photograph by the author)

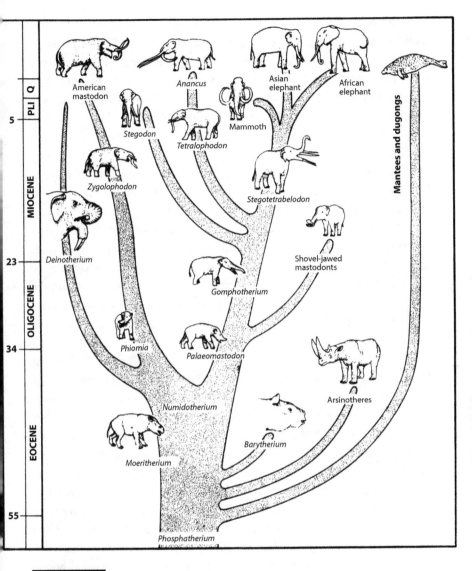

Figure 16.4 ▲

Evolutionary history of the elephants and their kin (Proboscidea), beginning with tapir-like or hippo-like forms such as *Moeritherium* with no trunk or tusks, through mastodonts with short trunks and tusks, and concluding with the huge mammoths and the two living species. Early in their history, the other tethytheres branched off from the Proboscidea. These include the manatees, order Sirenia, the extinct desmostylians, and the extinct horned arsinotheres. (Drawing by C. R. Prothero)

and flat forehead, with short tusks in both its upper and lower jaw. *Phiomia serridens* (whose name means "saw-toothed animal of the Fayûm") was a bit smaller, about 1.3 meters (4 feet) at the shoulder, with cylindrical upper tusks with an oval cross section, but lower tusks that were flattened into a spatula-like shape. Thus they are classic intermediate fossils in size between *Moeritherium* and later proboscideans in their tusks (longer than *Moeritherium* but shorter than later mastodonts) and in their trunk (intermediate in length as well).

The next split in proboscidean evolution was between the mastodon family (Mammutidae) and the rest of the Proboscidea. Mammutids can be traced back to *Eozygodon*, a very primitive fossil from the early Miocene of Kenya and Uganda. By the middle Miocene, they had evolved into *Zygolophodon*, which spread from Africa about 19 million years ago and, by 18 million years ago, had spread across Eurasia and even crossed the Bering land bridge to North America. But the best-known mammutid is the last of them all: the American mastodon, *Mammut americanum* (see figure 16.4). They were smaller than most living elephants, only about 3 meters (10 feet) at the shoulder, and about 4.5 tonnes (5 tons) in weight. *Mammut* lacked the steep forehead or shoulder hump found in elephants and mammoths. It had a long flat head with slightly curved tusks, a deeper chest, broader hips, shorter legs, and a longer back than mammoths as well. Its teeth retained the primitive condition of rounded conical cusps that connected into cross-crests as they were worn away. With these primitive teeth (and confirmed by the gut contents of mummified specimens), mastodons were mostly leaf, twig, and pine needle browsers that inhabited the forests of the Miocene, Pliocene, and Pleistocene, in contrast to the mostly grazing habits of mammoths and elephants. Mastodons had a thick coat of shaggy hair to keep them warm during the Ice Ages, but they were not as common or widespread as mammoths due to their habitat restrictions. They were thought to have become extinct with the rest of the Ice Age megamammals about 10,000 years ago, although there are legends of individual mammoths surviving into more recent times.

The main lineage of Proboscidea can be traced from *Palaeomastodon* of the early Oligocene to the gomphotheres of the early and middle Miocene. They were widespread across both North America and Eurasia during this time span, performing the role that mammoths and elephants later performed as the largest herbivore. Gomphotheres were about 3 meters (10 feet) at the shoulder and weighed 4 to 5 tons. They had long flat-topped skulls with two well-developed tusks in both their upper and lower jaws, and they probably had a short trunk. The lower tusks were shaped like spatulas and were thought to be useful for digging up roots and food, as well as for stripping bark off trees.

From the gomphotheres, many different groups of proboscideans evolved. One of these was the shovel-tuskers (subfamily Amebelodontinae, with five genera), whose name refers to the fact that the lower tusks are shaped like a pair of broad shovels. They evolved in North America about 9 million years ago, then spread to Asia in the late Miocene. Traditionally, shovel-tuskers were thought to have used their lower tusks for scooping up water plants in swampy habitats, but detailed analysis of the wear on their "shovels" shows abrasion from scraping bark off trees, so they probably had a diet of leaves and twigs and bark like most mastodonts. Shovel-tuskers vanished at the end of the Miocene in North America, the same time that rhinos, protoceratids, dromomerycines, musk deer, and many other groups typical of the American savanna vanished.

The final lineage of Proboscidea is the Elephantoidea, including the living elephants and the extinct mammoths. These lineages evolved in a new direction, with a tendency toward a shorter face and lower jaw and a raised crest on the top of the skull, which allowed them to develop the huge pair of upper tusks and lose the lower tusks entirely (figure 16.5). Meanwhile, their molar teeth became composed of a big set of tightly folded enamel and dentin plates that make a large grinding surface on the top, so they are well adapted for grinding tough vegetation like grasses. Eventually, these big teeth take over their short face and jaws, so they only had one or two molars in each side of the upper and lower jaw at the same time (figure 16.6A; figure 16.6B).

The final stage was the family Elephantidae, the modern family of elephants and mammoths. They can be traced back to the genus *Primelephas* ("first elephant" in Latin), a late Miocene genus that still had short upper and lower tusks like a gomphothere. During the Pliocene, they evolved into *Mammuthus*, the genus of mammoth, and spread across the Northern Hemisphere and Africa. As they evolved, their molars became larger and more complex, with more and more folds of enamel and dentin. Their evolution culminated with the huge Columbian mammoth, well known from temperate and tropical latitudes, and it had naked skin like a modern elephant. They reached 4 meters (13.1 feet) at the shoulder and weighed 7 to 9 tonnes (10 tons). But the fringes of the Pleistocene glaciers were inhabited by the woolly mammoths, which were slightly smaller and

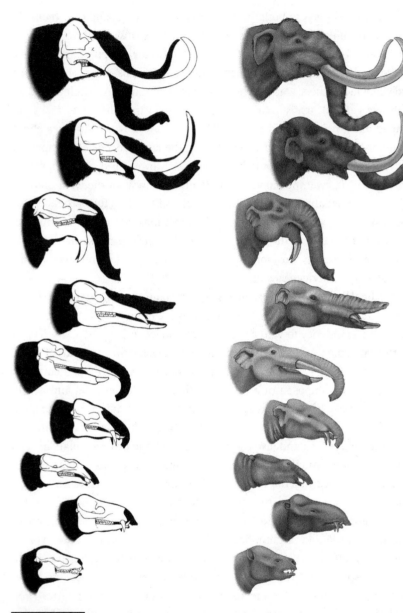

Figure 16.5 ▲

Details of the evolution of the skull, tusks, and trunk of proboscideans, from the pygmy hippo-like *Moeritherium* through mastodonts with longer tusks and trunks to mammoths. The genera are (*from bottom to top*): *Phosphatherium, Numidotherium, Moeritherium, Palaeomastodon, Phiomia, Gomphotherium, Deinotherium, Mammut* (American mastodon), and *Mammuthus* (the mammoth). (Illustration by Mary Persis Williams)

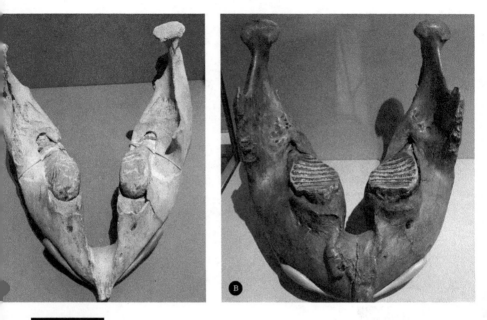

Figure 16.6 ▲

Mammoth molars in a jaw, (*A*) showing how they erupt from a crypt in the back of the jaw and then push forward, until (*B*) the worn molars break away from the front of the jaw. (Photographs by the author)

famously covered by a thick coat of long fur that protected them in their cold habitat. These mammoths are known from a number of mummified specimens (figure 16.7) frozen in the tundra of Siberia and Alaska, which show us not only what they looked like alive but also reveal what they ate (they seemed to have a fondness for buttercups).

Finally, at the end of the last Ice Age, about 10,000 years ago, all the mammoths vanished from the continents, along with most of the rest of the large Ice Age mammals. However, a dwarfed population of mammoths managed to survive in the Arctic islands of the Aleutians and near Siberia, persisting until about 6,000 years ago. There were also populations of dwarfed mammoths on other islands, such as the Channel Islands off the coast of Santa Barbara, California.

We have traced proboscidean evolution from Eocene forms like *Moeritherium* up to modern elephants. Where did *Moeritherium* come from? Recent discoveries in Africa have answered that question as well

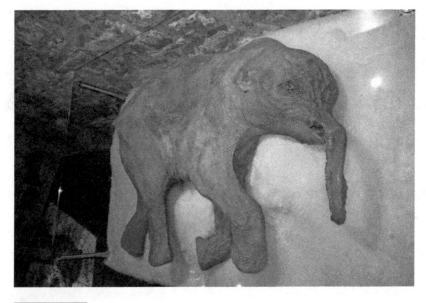

Figure 16.7 ▲
Freeze-dried mummified mammoth found in Siberian permafrost, with all the skin, hair, and most of the other soft tissues intact. (Photograph by the author)

(see figure 16.4; figure 16.5). The oldest known proboscidean relative was a fox-sized creature known as *Eritherium*, from the late Paleocene of Morocco, discovered in 2009. Its teeth have the four rounded cusps typical of primitive relatives of proboscideans, with just a hint of the cross-crest pattern seen in later proboscideans. The front teeth and lower jaw also show other features of the proboscideans, and the eyes are far forward on the skull, typical of the whole group. Also from the late Paleocene was *Phosphatherium*, a creature with many proboscidean features, including short snout bones but the beginning of a proboscis, yet its molars had better-developed cross-crests than those of *Eritherium*. Scratches on its teeth suggest that *Phosphatherium* ate a wide variety of plants, but mostly leaves. The next step is the early Eocene fossil *Daouitherium* from Morocco, and especially *Numidotherium* from Algeria. The incomplete skull of *Numidotherium* already shows signs of the tall, steep forehead so characteristic of elephants, with a nasal opening further back from the snout, suggesting a short trunk or proboscis. The upper canines are beginning to elongate into short tusks, but *Numidotherium* had not lost all of its front incisors as had later proboscideans. Their lower front teeth are beginning to form a scoop shape typical of early mastodons. It was about 1 meter (3 feet) tall at the shoulder, but it had robust limbs typical of all proboscideans, even though it was only pig-sized.

The evolution of the proboscideans is a splendidly documented fossil sequence that shows the complex branching history of this group from fox-sized creatures with only a few elephant-like features, gradually enlarging their trunks and sporting a variety of upper and lower tusks, until we reach modern elephants and extinct mammoths. How the elephant got its trunk is no longer a mystery; it is one of the best documented stories in the fossil record.

Finally, we have one more pressing question: Where did the proboscideans come from? What is their nearest relative? In 1975, Malcolm McKenna, my graduate advisor at the American Museum of Natural History in New York, proposed a group called Tethytheria. They were a group of mammals whose origins apparently occurred around the edges of the tropical Tethys Seaway that once stretched from Gibraltar to Indonesia. Originally, the tethytheres included only the proboscideans and another well-known group, the sirenians or sea cows (familiar to most people from manatees and dugongs). As McKenna pointed out, they had a large number of unique anatomical features in their skull, including the tendency to replace their baby teeth not with adult teeth pushing up from below but with new teeth pushing from the back of the jaw, with old worn teeth falling out at the front of the jaw. Only elephants and manatees do this today, and only their extinct relatives show this peculiar pattern known as horizontal tooth replacement. In the 1980s and 1990s, molecular studies confirmed that elephants and manatees were each other's closest living relatives.

Since that time, further anatomical analysis of fossils have shown that the tethytheres also include two extinct groups: the weird horned arsinoitheres, found by Andrews and Beadnell in the Fayûm Depression in 1902 (see figure 16.4), and a strange aquatic group of hippo-like mammals known as desmostylians, found only in the Miocene marine rocks of the Pacific Rim from Baja California to Japan. The last surprise from molecular biology showed that tethytheres were closely related to a bunch of African groups, including the woodchuck-like hyraxes, the termite-eating aardvarks, plus the tiny elephant shrews, the shrew-like insectivores from Madagascar known as tenrecs, and the strange mole-like animals known as the golden moles. All of them either live entirely in Africa today, or originated there and spread elsewhere, so they have been called the Afrotheria. Their zoological affinities were long a mystery until the evidence from molecular biology came down strongly in favor of them all being closely related.

A few mammoth bones and an isolated molar from the American *incognitum* gave us the first idea about extinct elephant relatives, but we now have a rich fossil record tracing their history to their earliest forms more than 60 million years ago. In addition, the fossil record plus molecular biology unites them with a whole bunch of odd animals, including manatees, aardvarks, hyraxes, tenrecs, and golden moles. That is a story no one could have imagined even 50 years ago.

FOR FURTHER READING

Gheerbrant, Emmanuel. "Paleocene Emergence of Elephant Relatives and the Rapid Radiation of African Ungulates." *Proceedings of the National Academy of Sciences* 106, no. 26 (2009): 10717–10721.

——. "A Palaeocene Proboscidean from Morocco." *Nature* 383, no. 6595 (1996): 68–70.

Prothero, Donald R., *The Princeton Field Guide to Prehistoric Mammals*: Princeton, NJ: Princeton University Press, 2016.

Prothero, Donald R., and Robert M. Schoch. *Horns, Tusks, and Flippers: The Evolution of Hoofed Mammals*. Baltimore, Md.: Johns Hopkins University Press, 2002.

Rose, Kenneth D., and J. David Archibald, eds. *The Rise of Placental Mammals: Origins and Relationships of the Major Extant Clades*. Baltimore, Md.: Johns Hopkins University Press, 2005.

Savage, Robert J. G., and M. R. Long, *Mammal Evolution: An Illustrated Guide*. New York: Facts-on-File, 1986.

Shoshani, Jeheskel. *Elephants: Majestic Creatures of the Wild*. New York: Rodale Press, 1992.

Shoshani, Jeheskel, and Pascal Tassy. "Advances in Proboscidean Taxonomy and Classification, Anatomy and Physiology, and Ecology and Behavior." *Quaternary International* 126–128 (2005): 5–20. https://doi.org/10.1016/j.quaint.2004.04.011.

——, eds. *The Proboscidea: Evolution and Paleontology of Elephants and Their Relatives*. New York: Oxford University Press, 1996.

Turner, Alan, and Mauricio Anton. *National Geographic Prehistoric Mammals*. Washington, D.C.: National Geographic Society, 2004.

PART IV

EYES AND GENES

A WARM LITTLE POND

It is often said that all the conditions for the first production of a living organism are now present, which could ever have been present. But if (and oh! what a big if!) we could conceive in some warm little pond, with all sorts of ammonia and phosphoric salts, light, heat, electricity, &c., present, that a proteine [sic] compound was chemically formed ready to undergo still more complex changes, at the present day such matter would be instantly absorbed, which would not have been the case before living creatures were found.

—CHARLES DARWIN, IN AN 1871 LETTER TO JOSEPH HOOKER

The origin of life is one of the most commonly misunderstood and distorted topics that comes up when evolution is discussed in the public arena. Here are some commonly voiced lies and misconceptions about the origin of life:

"I can't imagine life arising from random chance."

"The probability that life can arise from nonlife is so small that it is like a tornado blowing through a junkyard and assembling a 707."

"Science shows that spontaneous generation cannot occur, so how does life arise?"

These and other mistaken and misleading arguments are heard all the time, and they have been answered many times as well. Despite debunking these ideas, scientific explanations for the origin of life never seem to register in the public consciousness.

Strictly speaking, explaining the origin of life (abiogenesis) is not part of classic evolutionary theory at all, which primarily deals with natural selection once life had originated. But it's such an easy way for science deniers to exploit the personal incredulity and lack of science literacy in their audience that it comes up all the time.

Recent breakthroughs in origin of life research have produced remarkable discoveries, many of which defy explanation by those who deny the reality of evolution. For example, if specially created, why do mitochondria and chloroplasts in the eukaryotic cell have their own genomes and can be killed by antibiotics? Evolution has a well-tested explanation for that. Why is all life based on only 20 out of more than 100 known amino acids, with only one type of mirror-image symmetry (right handed)? Evolution also explains that beautifully. If life were specially created, why are the building blocks of life (amino acids) found all over the universe? Again, this only makes sense in light of evolution. Many remarkable discoveries in the lab have successfully replicated nearly every step in the origin of life, and we are very close to creating life from nonlife in the lab as this book is being written.

First, let's address the misconceptions about spontaneous generation. Before the 1860s, most people (including Lamarck and Darwin) believed that life could spontaneously arise from nonlife. Maggots mysteriously appeared in rotting meat, and broth left out would spoil as bacteria and fungi took over. Then Louis Pasteur did a famous series of experiments that showed that spontaneous generation does not occur—maggots or bacteria only grow if there is some way for them to propagate from living organisms outside the experiment. But all those experiments assumed present Earth conditions, especially the present-day atmospheric oxygen levels of 21 percent. In fact, overwhelming geologic evidence shows that Earth's early atmosphere had no free oxygen when it formed 4.6 billion years ago, and only about 1 percent of the atmosphere had oxygen 2.3 billion years ago. In those conditions, it is much easier to make life from nonlife. In fact, judging from the presence of biotic chemicals such as amino acids all over the universe, it is much easier to create life than one would expect. Once life did arise, it gobbled up all of the rich nutrients so no other creatures could exploit them, thus preventing life from originating again. So that's the important caveat: *Under present Earth conditions with free oxygen, spontaneous generation does not occur.*

Second, how could random chance assemble something as complex as life? Once again, the premise of the argument is completely wrong. It is true that genetic variation occurs due to chance mutations and recombination, but *evolution does not occur by chance*—natural selection is a nonrandom process, weeding out the less fit from the fit. An old analogy suggests that the probability that a monkey randomly hitting keys on a typewriter will somehow write the works of Shakespeare is extremely small. (This analogy is equally dated because typewriters are nearly extinct.) A much better analogy can be made for a monkey with a spell-checker in his word processing software. The spell-checker automatically fixes and corrects mistakes when it recognizes a combination that might be functional, and it eliminates many of the nonsense streams of characters generated by randomly striking keys. Natural selection is like a nonrandom spell-checker, getting rid of bad combinations arising by chance and favoring those that are functional. If you perform some simple computer simulations of this process, you will find that a recognizable string of meaningful words can be produced in just a few dozen to a few hundred iterations of the routine— because the software is editing out the mistakes and selecting only for combinations that work. In *The Blind Watchmaker* (1986) and *Climbing Mount Improbable* (1996), Richard Dawkins provides many interesting examples and computer models that show just how easily this can be done.

Other false analogies have been put forward that assume random chance is operating. Astronomer Fred Hoyle (later copied by Duane Gish) coined the famous false analogy of a tornado in a junkyard assembling a 707 (again, a highly dated analogy starring a plane from 60 years ago). But evolution is not a random destructive force like a tornado or hurricane; it is a slow, methodical, tinkering, spell-checking computer program that builds order out of random ingredients through programming that selects for combinations that work.

Very often, science deniers will say, "How could the many steps needed to make a complex cell happen by random chance? The probability against it is enormous!" They will then proceed to do a nonsensical calculation of probability based on the false assumption that everything is operating by chance. As anyone who understands probability knows, you can't make this kind of argument after the fact. If you do so, any complex sequence of events is extremely improbable, even though some actually occur. During a debate, I once asked the audience of several hundred to estimate the probability *after the fact* that all of the events that had happened in their lives would actually happen, and the probability that among all those unlikely events, they would all end up in this room at this particular moment.

The improbability of this event is, of course, enormous. Using my debate opponent's probability arguments, I pointed out that this audience could not exist! So let's set aside these false analogies and mistaken arguments and look at the positive scientific research on how life originated.

The scientific pursuit of the problem of the origin of life goes all the way back to Darwin, who suggested a model in an 1871 letter to his friend, botanist Joseph Hooker (see the chapter epigraph). Darwin speculated that a "warm little pond" with the right combination of chemical compounds (whimsically nicknamed the "primordial soup" by later scientists) and the right sources of energy could produce proteins. In the 1920s, Russian biochemist A. I. Oparin and British geneticist J. B. S. Haldane both came up with the idea that Earth originally had a reducing atmosphere of nitrogen, carbon dioxide, ammonia (NH_3), and methane or natural gas (CH_4), but no free oxygen. Such an atmosphere and corresponding ocean would be the ideal primordial soup for producing simple organic compounds.

In 1953, Stanley Miller, a young graduate student at the University of Chicago, in collaboration with his advisor chemist Harold Urey, decided to test Oparin's hypothesis. Urey later won the Nobel Prize in Chemistry for his many achievements in isotopic chemistry, including the discovery of deuterium and the method of separation of isotopes needed to make nuclear bombs in the Manhattan Project. Miller wanted to see if such a primordial soup could generate biochemicals. He built a simple apparatus out of glass tubing that formed a continuous sealed loop, with all the air (including oxygen) removed by vacuum pump. He supplied a new "atmosphere" rich in carbon dioxide, nitrogen, methane, ammonia, and water (but *no* free oxygen) in the evacuated tubes. Heating the "ocean" flask at the base to start the steam circulating, he used sparks in another flask to simulate "lightning" as an energy source (figure 17.1). Once the methane-ammonia-laden steam moved through the "lightning," a condenser cooled the steam, and the liquid water flowed back to the "ocean" flask.

This simple experiment produced amazing results. Within days, the clear solution of the "ocean" became yellowish-brown with new chemicals, and within a week, it was a dark brown organic-rich glop. When Miller analyzed it, he found that he had already produced 4 of the 20 amino acids used to make proteins, plus many other simple but crucial organic molecules, such as cyanide (HCN) and formaldehyde (H_2CO). With one remarkable experiment, Miller had launched the whole field of biochemical research into the origin of life.

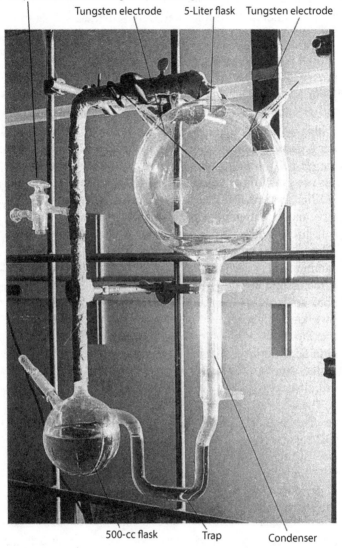

Stopcock for withdrawing samples during run

Tungsten electrode   5-Liter flask   Tungsten electrode

500-cc flask   Trap   Condenser

Figure 17.1 ▲

An apparatus like this was used by Stanley Miller and Harold Urey in 1953 to simulate the synthesis of complex organic compounds on the early Earth. Air was evacuated from the system and the large flask held an "atmosphere" rich in carbon dioxide, water, nitrogen, ammonia, and methane (but no oxygen). Sparks from electrodes simulated lightning. The product of this reaction then flowed through the condenser and accumulated in the flask, which became a brew of "primordial soup." After about a week, the clear solution had turned into a thick murky brown sludge full of newly synthesized organic compounds, including many of the amino acids necessary to build life. (Courtesy of Stanley Miller)

Although amino acids are much more complex than the chemicals he started with, Miller showed that they were remarkably easy to produce. Other labs later conducted experiments similar to those performed by Miller and produced 12 of the 20 amino acids found in life. Experimenting with a dilute cyanide mixture produced 7 amino acids. No matter what experiment you devise, it does not require supernatural powers or even more than a few days in the lab to make the basic building blocks of life. In fact, Miller's experiment is so simple that anyone with access to a decent chemistry lab and a vacuum pump and the right gases can do it, and online articles describe how to set up your own Miller-Urey experiment. In the years since Miller's original experiments, other scientists have found 74 different amino acids trapped in meteorites that were formed in the original solar system (including all 20 of those found in living systems). Amino acids are remarkably easy to produce, so we can assume that they were present in Earth's early oceans (as they were in certain meteorites and apparently throughout space).

But we are interested in the more complex biochemicals known as proteins, which are made of long chains of amino acids. Proteins are the fundamental building blocks of most living systems. For a true living organism, we also need other complex chains composed of simple building blocks. We need lipids (common in oils and fats), built out of a combination of a long chain of fatty acids linked to alcohols. We need to assemble carbohydrates and starches, which are composed of long chains of simple sugars such as glucose. And we need nucleic acids (RNA and DNA), composed of complex chains of sugars, phosphates, and the four bases (adenine, thymine, cytosine, and guanine), which carry the genetic code necessary for making copies of the organism. All these complex molecules are polymers, and they are formed by linking together simpler components (figure 17.2), a reaction called polymerization. How do we trigger these polymerization reactions?

It turns out that many of these reactions are easy to produce and readily make short-chain polymers. In the 1950s, Sidney Fox splashed a solution of amino acids on hot dry volcanic rocks, and this mixture formed many of the proteins found in life. In the presence of formaldehyde, certain sugars readily form complex carbohydrates. Miller's early experiments also produced the components of nucleic acids, such as the nucleotide base adenine (by heating aqueous solutions of cyanide) and adenine plus guanine (by bombarding dilute hydrogen cyanide with ultraviolet radiation).

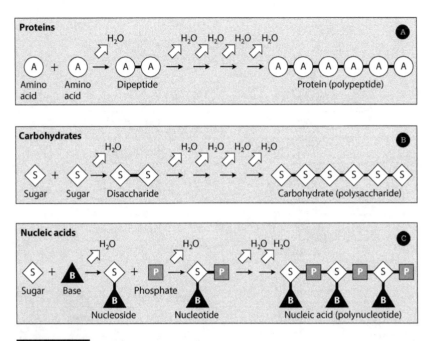

Figure 17.2 ▲
The next step in the origin of life is arranging the smaller building blocks into longer, more complex chains (polymerization). The common reactions include linking together numerous amino acids to form proteins, the basic building blocks of life; polymerizing simple sugars into complex carbohydrates, the basic component of cell walls and also a critical energy source in metabolism; and linking together sugars, phosphates, and nucleosides to make nucleic acids (DNA and RNA), the basic genetic code of all life. (Courtesy of J. William Schopf)

Lipids are even easier to produce (figure 17.3). Their basic building blocks are a water-soluble alcohol, glycerol, at one end, and a long fatty acid chain sticking out at the other end. The alcohol end is hydrophilic (soluble in water), and the fatty acid end is hydrophobic (does not dissolve in water). When you put some lipids in water, the hydrophilic end of the molecule orients toward the water, and the hydrophobic fatty acid points away from the water. With enough lipids, you can generate a lipid bilayer, with all the lipids lined up and closely packed together in the same orientation, allowing them to link together. You see this reaction any time you put a droplet of oil in water, or a droplet of water in oil. As everyone knows, oil and water don't mix—they form discrete fluid masses separated by lipid bilayers. It turns out that the simple membranes of most primitive organisms are also lipid

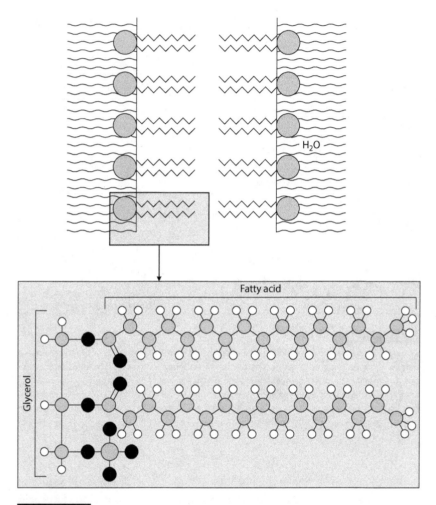

Figure 17.3 ▲

Some organic chemicals have properties that enable cells to form naturally without complex organic reactions. Lipids, the building blocks of fats and oils, have an end that repels water and an end that bonds to water. The properties that repel or attract lipids to water naturally line up and then combine to form membranes. Whenever oil mixes with water, it forms a natural membrane that encloses a droplet, comparable to the lipid bilayer membrane that surrounds all cells. (Courtesy of J. William Schopf)

bilayers. So generating a membrane for our most primitive living organism is as simple as making oil and water salad dressing!

When these lipid droplets are dried and then wetted again, they form spherical balls that concentrate any DNA present up to 100 times. These little lipid bilayer droplets with nucleic acids trapped inside have all the properties of "protolife." Oparin produced droplets he called "coacervates," and Sidney Fox produced similar structures that he called "proteinoids." These droplets behave much like living cells, holding together when conditions change, growing, and budding spontaneously into daughter droplets. They selectively absorb and release certain compounds in a process similar to bacterial feeding and excretion of waste products. Some even metabolize starch! Even though these protocells are not living, they have most of the properties of living cells—all created using simple chemical reactions plus heat.

It is relatively easy to form the basic building blocks of life in an organic chemistry lab—amino acids and short-chain proteins; simple sugars and starches; fatty acids plus alcohol to make lipid bilayers and cell membranes; and short nucleic acids to pass on the genetic information—but most living systems are built from molecules that are many hundreds to thousands of units long. It is difficult to assemble these molecules when the individual building blocks are randomly bumping into each other in a solution, but there is a more efficient, natural way to bring these molecules closer together in the right pattern to link up into complex molecules. Organic chemists often use a catalyst, some sort of inorganic substance added to the solution, to speed up the reaction. In nature, many such catalysts could serve to line up the building blocks of life into a tightly packed framework of molecules. These catalysts can be thought of as "templates" or "scaffolds," an external inorganic framework that holds the smaller organic molecules in position until they are all lined up in the right direction and jostling against one another. It is analogous to the arrangement of people in a mosh pit, packed in shoulder to shoulder and all oriented facing the stage. If they are closely packed enough, their large earrings and other piercings might link together, and they would be assembled into a tightly linked chain of people. Likewise, if you pack organic molecules in the proper orientation and very closely together, their "earrings" ($OH-$ and $H+$, or hydroxyls and hydrogens, sticking out at the end of each chain) connect (see figure 17.2), leaving two larger molecules linked together.

What kinds of scaffolds or templates might be able to catalyze such reactions? The most promising is fool's gold or, more precisely, the mineral known as pyrite (iron sulfide, FeS_2). Pyrite has a metallic golden appearance, hence the nickname fool's gold. But pyrite is an important mineral in certain settings, especially in highly reducing fluids depleted in oxygen in which the iron combines with sulfur rather than with oxygen. Iron sulfides are common in deep anoxic oceanic waters such as the bottom of the Black Sea, where the reducing conditions form FeS_2 rather than iron oxides, which are typical of black shales. Pyrite is especially abundant in the midocean ridge volcanic vents, or "black smokers" (figure 17.4), where dissolved sulfides from the oceanic crust rise with the superheated water from the magma chamber of the midocean ridge. This is intriguing because a number of scientists have argued that the chemistry of these sulfide vents is ideal for the production of earliest life. Not only is there plenty of energy in the form of volcanic heat, but several chemists have argued that pyrite is also a good scaffold or template. Its crystal surfaces are electrically charged, so organic molecules are naturally attracted to it and are attached by their oppositely charged ends. When they become a densely packed mosh pit, they will link together by the condensation reaction and form long-chain biochemicals on the pyrite scaffold.

The midocean ridge vent theory for life's origins has another advantage—it is isolated from most events at the earth's surface and very stable. Even when meteorites pounded the earth between 4.6 and 3.9 billion years ago, vaporizing the shallow oceans over and over again, the deep ocean vents were protected. And the final interesting convergence of lines of evidence is that the most primitive organisms on Earth, the Archaebacteria or Archaea, are found in these same extreme settings with boiling waters and an abundance of sulfur.

There is no shortage of good mechanisms to naturally assemble small organic molecules into the long-chain biochemicals that life requires. Some of these proposed templates also fit with an increasing body of evidence suggesting that life originated in the deep-sea volcanic vents, not in Darwin's "warm little pond" as was long supposed.

I have shown that the origins of the basic biochemicals (amino acids and proteins, sugars and carbohydrates, lipids and cell membranes, plus nucleic acids) needed for life are easily produced by simple natural chemical reactions, and some even occur in space. These simple short-chain polymers

Figure 17.4 ▲

In the deep volcanic rift valleys of midocean ridges, fresh lava erupts as the oceanic crust pulls apart. The magma heats the seawater percolating through it to superheated temperatures, forming plumes of boiling water and dissolved minerals known as "black smokers." The main precipitate of this reaction is pyrite (iron sulfide, or fool's gold), which is also a good template for bonding complex organic materials. Consistent with the hypothesis that life originated in deep-sea vents, biologists have found that the genetically simplest forms of life, the archaebacteria, are common in the black smokers. These are the base of a food chain that includes a huge community of giant clams, tube worms, crabs, and many other unique creatures found only in these dark submarine communities. There is no light at this depth, so the entire system relies not on photosynthesis but on chemosynthesis, with sulfur-reducing bacteria (rather than plants) at the base of the food chain. (Courtesy of NOAA)

are naturally and easily linked together into the long-chain polymers by catalysis on scaffolds of some nonorganic matrix, such as pyrite. All of these steps give us a nucleic acid wrapped by a lipid bilayer coat, which has many cell-like metabolic functions. In short, this hypothetical earliest life form is not too different from the most primitive bacterial cells, the simplest known organisms, which are not much more than a nucleic acid wrapped in a cell membrane with other added functions.

Bacteria and other very simple organisms are prokaryotes, organisms that have their nucleic acid genes (either RNA or DNA) floating in the cell without a nucleus. These tend to be very small (only a few microns in diameter) and very simple. The earliest fossils known (from rocks over 3.4 billion years old in Australia and South Africa) are of the size and shape that we can confidently attribute them to prokaryotes, including blue-green bacteria (cyanobacteria, once known incorrectly as "blue-green algae") and other types of bacteria. But more complex organisms are known as eukaryotes, and they have a type of eukaryotic cell (found in animals, plants, and fungi) that is about 10 times larger than a prokaryotic cell (figure 17.5). All of the nucleic acid genes (DNA) are enclosed in a membrane surrounding a nucleus, and eukaryotic cells almost always have additional structures (organelles) within the cell wall. These might include chloroplasts, which are the sites of photosynthesis in plant cells; mitochondria, the "power plants" of the cell where energy is exchanged using ATP and ADP; Golgi bodies, which process and package proteins; endoplasmic reticulum, which synthesize proteins, lipids, steroids, and

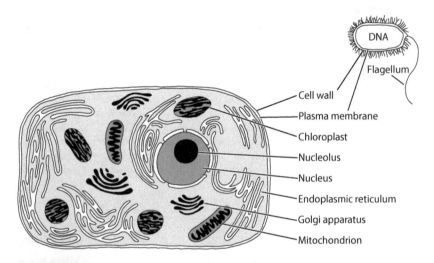

Figure 17.5 ▲

Prokaryotes, such as the archaebacteria and true bacteria, are small cells only a few microns in diameter. Their genetic material (DNA) is not enclosed within a nucleus but floats within the cell, and they lack organelles. Eukaryotes (all other living organisms) have larger, more complex cells, with a discrete nucleus containing their DNA. They also may have a number of other organelles, including mitochondria, chloroplasts, Golgi apparatus, endoplasmic reticulum, cilia, flagella, and other subcellular structures. (From Donald Prothero, *Evolution: What the Fossils Say and Why It Matters*, 2nd ed. [New York: Columbia University Press, 2017])

other chemicals, and also regulate concentration of calcium and other steroids; and external structures, such as the hair-like cilia used in propulsion, or the whip-like flagellum used to power the cell rapidly through a fluid. For a long time, how all of these complex structures had evolved from scratch was a great puzzle.

In 1967, biologist Lynn Margulis proposed a radical solution to this problem (Russian botanist Konstantin Mereschowski had proposed a version of this idea in 1905, but it was then forgotten). Instead of the difficult process of evolving organelles out of nothing, Margulis argued that each of the organelles found in the eukaryotic cell were once free-living prokaryotes that had come to live symbiotically within another cell and eventually had become part of it (figure 17.6). This idea is known as the endosymbiosis theory.

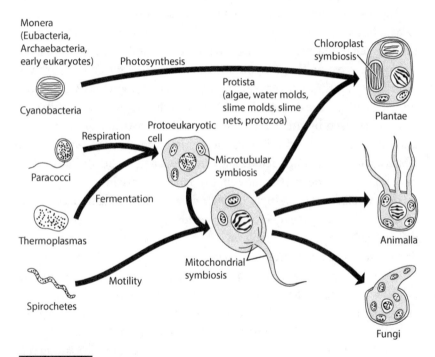

Figure 17.6 ▲

According to Lynn Margulis, complex eukaryotic cells arose from two or more prokaryotic cells that combined to live symbiotically. Cyanobacteria are apparently the precursors of the photosynthetic chloroplasts, which provide photosynthesis in plant cells. Purple nonsulfur bacteria have the same structure and genetic code as mitochondria, which provide energy in the cell. And the flagellum has the same structure as the prokaryotes known as spirochetes, which are also responsible for causing syphilis. (From Donald Prothero, *Evolution: What the Fossils Say and Why It Matters*, 2nd ed. [New York: Columbia University Press, 2017])

Chloroplasts apparently started out as cyanobacteria, which are photosynthetic even though they are prokaryotes without organelles. Purple nonsulfur bacteria have much the same structure and function as mitochondria, and apparently that was the origin of these organelles. The flagellum has the identical 9 + 2 fiber structure (9 sets of microtubule doublets surrounding a pair of single microtubules in the center) as the prokaryotes known as spirochaetes, which also cause the disease known as syphilis. As each of these smaller prokaryotes came to live within a larger cell, they sublimated their functions to that of their host; the cyanobacteria became chloroplasts that are now homes for photosynthesis, and the purple nonsulfur bacteria became mitochondria and performed the role of the energy converter for the cell.

In addition to the detailed similarities of these prokaryotes to the organelles, Margulis pointed to many other suggestive lines of evidence. Organelles are not usually floating within the eukaryotic cell membrane but are separated from the rest of the cell by their own membranes, strongly suggesting that they are foreign bodies that have been partially incorporated within a larger cell. Mitochondria and chloroplasts also make proteins with their own set of biochemical pathways, which are different from those used by the rest of the cell. Chloroplasts and mitochondria are also susceptible to antibiotics such as streptomycin and tetracycline, which are good at killing bacteria and other prokaryotes, but the antibiotics have no effect on the rest of the cell. Even more surprising, mitochondria and chloroplasts can multiply only by dividing into daughter cells like prokaryotes. They have their own independent reproductive mechanisms and are not made by the cytoplasm of the cell. If a cell loses its mitochondria or chloroplasts, it cannot make more.

When Margulis's startling ideas were first proposed more than 50 years ago, they were met with much resistance. But as biologists began to see more and more examples of symbiosis in nature, the notion became more plausible. We humans have millions of endosymbionts (mostly bacteria) on our skin and inside us. Our intestines are full of the bacterium *Escherischia coli* (*E. coli* for short), familiar from petri dishes and news alerts about sewage spills or contaminated kitchens. These bacteria do most of our digestion for us, breaking down food into nutrients in exchange for a home in our guts. Most of our fecal matter consists of the dead bacterial tissues after digestion, plus indigestible fiber and other material that we cannot metabolize.

There are many other examples of endosymbiosis in nature. Termites, sea turtles, cattle, deer, and goats and many other organisms have specialized gut bacteria that help break down indigestible cellulose, enabling these animals can eat plant matter efficiently. Tropical reef corals and giant clams all house symbiotic algae in their tissues, which produce oxygen, remove carbon dioxide, and help secrete the minerals for their large shells.

The strongest evidence came when people started studying the organelles more closely and found that not only did they have the right structure to have once been independent prokaryotic cells but they also *have their own genetic code!* Mitochondria and chloroplasts both have their own DNA, which have a different sequence than the DNA found in the cell nucleus. In fact, mitochondrial DNA evolves at a different rate from nuclear DNA and is different enough that it can be used to solve problems of evolution that nuclear DNA cannot. This would make no sense if the eukaryotic cell had tried to generate the organelles from scratch. They would not have their own genetic code if that were true. This discovery is one of the most powerful lines of evidence that life evolved and was not created.

The clincher is that many living endosymbiotic cells show that this process is occurring right now. The simpler eukaryotes, such as the freshwater amoebas *Pelomyxa* and the *Giardia* (famous for causing dysentery from contaminated water), lack mitochondria but contain symbiotic bacteria that perform the same respiratory function. In the laboratory, scientists have observed amoebae that have incorporated certain bacteria in their tissues as endosymbionts. The parabasalids, which live in the guts of termites, use spirochetes for a motility organ instead of a flagellum. From wild speculation in 1967, Margulis's idea has now come to be accepted as the best possible explanation of the origin of eukaryotes and organelles.

So, contrary to popular belief, the origin of life by natural processes is not improbable nor difficult to imagine. In fact, nearly all the steps have been simulated in the laboratory, from the synthesis of amino acids (found all over the universe); to the polymerization of simple biomolecules into long-chain building blocks of life such as proteins, lipids, carbohydrates, and nucleic acids; to even more complex biochemical organisms built on a charged surface of pyrite in deep-sea black smoker hydrothermal events, where the most primitive forms of life (archaebacteria) are found even today. Then we learned of the assembly of complex eukaryotic cells with the endosymbiotic prokaryotic cells becoming their organelles. There is strong evidence (direct or indirect) that all of these stages actually happened, and most stages have been simulated in the laboratory.

In recent years, numerous researchers have seen novel genes evolve in the test tube, new genes assemble themselves into more complex life forms, and even multicellular life arising from single-celled life in a test tube. In one experiment, scientists observed two different lineages of RNA evolving and changing and even competing with one another. We have not yet produced life itself in a test tube, but the origin of life is not that mysterious, nor does it remain unsolved. The steps are all well-known, and laboratory results have confirmed each step along the way.

FOR FURTHER READING

Cone, Joseph. *Fire Under the Sea: The Discovery of the Most Extraordinary Environment on Earth—Volcanic Hot Springs on the Ocean Floor.* New York: Morrow, 1991.

Fry, Iris. *The Emergence of Life on Earth: A Historical and Scientific Overview.* New Brunswick, N.J.: Rutgers University Press, 2000.

Knoll, Andrew H. *Life on a Young Planet: The First Three Billion Years of Evolution on Earth.* Princeton, N.J.: Princeton University Press, 2003.

Knoll, Andrew H., and Sean B. Carroll. "Early Animal Evolution: Emerging View from Comparative Biology and Geology." *Science* 284 (1999): 2129–2137.

Hazen, Robert M. *Gen-e-sis: The Scientific Quest for Life's Origins.* New York: Joseph Henry Press, 2005.

Margulis, Lynn. *Symbiosis in Cell Evolution.* San Francisco, Calif.: W. H. Freeman, 1981.

——. *Symbiotic Planet: A New Look at Evolution.* New York: Basic Books, 2000.

Miller, Stanley L. "A Production of Amino Acids Under Possible Primitive Earth Conditions." *Science* 117, no. 3046 (1953): 528–529.

Schopf, J. William. *Cradle of Life: The Discovery of Earth's Earliest Fossils.* Princeton, N.J.: Princeton University Press, 1999.

——. *Life's Origin: The Beginnings of Biological Evolution.* Berkeley: University of California Press, 2002.

Shapiro, Robert. *Origins: A Skeptic's Guide to the Creation of Life on Earth.* New York: Summit, 1986.

Wills, Christopher, and Jeffrey Bada. *The Spark of Life: Darwin and the Primeval Soup.* New York: Perseus, 2000.

GENETIC JUNKYARD

It may be said that natural selection is daily and hourly scrutinising, throughout the world, every variation, even the slightest; rejecting that which is bad, preserving and adding up all that is good; silently and insensibly working, whenever and wherever opportunity offers, at the improvement of each organic being in relation to its organic and inorganic conditions of life. We see nothing of these slow changes in progress, until the hand of time has marked the long lapses of ages, and then so imperfect is our view into long past geological ages.

—CHARLES DARWIN, *ON THE ORIGIN OF SPECIES* (1859)

The year 1953 was a landmark in the history of science. At the University of Chicago, Stanley Miller conducted the first experiments in the field of origin of life (see chapter 17). At the Lamont-Doherty Geologic Observatory of Columbia University, Maurice "Doc" Ewing, Bruce Heezen, and Marie Tharp discovered and mapped the rift valley in the mid-Atlantic ridge, the first indication that seafloor spreading is occurring between oceanic crustal plates, which became the foundation of plate tectonics a decade later. Alan Turing published one of the most influential papers in computer science. Jonas Salk announced the first polio vaccine. The "Piltdown Man" skull was finally revealed to be a hoax. Frederick Reines and Clive Cowan reported the first successful detection of neutrinos. And in a lab at Cambridge University, two young scientists, Francis Crick and Jim Watson, figured out the structure of the DNA molecule and confirmed that it was the "code" that all life used to construct their bodies and to make copies that were passed on to the next generation.

After the first flush of confirming that DNA was the blueprint of life, and deciphering its detailed structure and the mechanism of the copying and translation of DNA, genetics began to move on to an even larger topic: finding the genetic code itself. Early studies in DNA showed that the double helix was an external "backbone" made of phosphates and sugars (like ribose or deoxyribose) and arranged as a spirally twisted "ladder" (figure 18.1). The "rungs" of the ladder were a series of nitrogenous bases known as adenine, thymine, guanine, and cytosine (A, T, G, and C). (In RNA, uracil replaces thymine.) Combined with the sugar and phosphate of the ladder, this unit is known as a nucleotide.

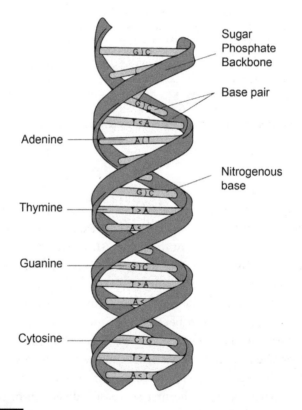

Figure 18.1 ▲

The structure of the DNA molecule. Sugars and phosphates form the double-helix "backbone" of the molecule, and the different bases (adenine, thymine, guanine, cytosine) form the "rungs" of the ladder, linking together with their matching base to create a "code" with their sequence of bases.

As early as 1959, a group led by Francis Crick found that a sequence of three nucleotides in a row (called a codon) was all that was necessary to code for a protein. Then Marshall Nierenberg and Heinrich Matthaei of the National Institutes of Health used a clever technique to figure out which three nucleotides in a codon spelled out which protein. They synthesized an RNA strand of nothing but uracils (UUUUUU. . . .) and produced only the protein phenylalanine. They presented their results at the International Congress of Biochemistry in Moscow in 1961. Francis Crick was so impressed that he persuaded the conference to listen to Nierenberg's talk again the next day to the entire congress. They then showed that a sequence of nothing but adenines (AAAAA. . . .) produced the protein lysine, and nothing by cytosines (CCCCCC. . . .) produced the protein proline. Soon all of molecular biology was focusing on the coding race, with several labs competing to see who could decipher the genetic code first.

Severo Ochoa's lab at New York University (with its large staff) was leading in the race. Nierenberg's small NIH lab could not compete, so many NIH scientists laid down their own research to help him sequence as much of the code as possible. DeWitt Stetten, the lab's director, called it the "NIH's finest hour." Finally, in the early 1960s, Har Gobind Khorana of the University of British Columbia deciphered the rest of the genetic code. The discovery was so momentous that Nierenberg, Khorana, and R. W. Holley (who discovered transfer RNA that reads the genetic code) shared the 1968 Nobel Prize in Physiology or Medicine.

Once the genetic code was deciphered, the shocking thing about it was its redundancy. Of the three-base sequence, all that usually mattered were the first two "letters" (= bases) in the code (figure 18.2). The third base (letter) is typically redundant and does not change which amino acid is produced. For example, any sequence that begins GU . . . produces valine, whereas AC . . . produces threonine, and CG . . . produces arginine. Only a few of the codes require the third letter to specify which amino acid they produce, and even then there are usually two possible options (for example, CAU and CAC produce histidine, and AAA and AAG both produce lysine). These 64 possible combinations of three letters specify only the 20 amino acids needed for life, plus a few "stop" codes (to end transcription) and a "start" code (to begin transcription).

In 1962, a few scientists noticed this redundancy and realized that mutations in the third (silent) position in the codon would be invisible to natural

The genetic code, which specifies by three letters in the genome (A = adenine; C = cytosine; G = guanine; U = uracil) any one of 20 amino acids, or a stop command.

First base in the codon	Second base in the codon				Third base in the codon
	U	**C**	**A**	**G**	
U	Phenylalanine	Serine	Tyrosine	Cysteine	U
	Phenylalanine	Serine	Tyrosine	Cysteine	C
	Leucine	Serine	Stop	Stop	A
	Leucine	Serine	Stop	Tryptophan	G
C	Leucine	Proline	Histidine	Arginine	U
	Leucine	Proline	Histidine	Arginine	C
	Leucine	Proline	Glutamine	Arginine	A
	Leucine	Proline	Glutamine	Arginine	G
A	Isoleucine	Threonine	Asparagine	Serine	U
	Isoleucine	Threonine	Asparagine	Serine	C
	Isoleucine	Threonine	Lysine	Arginine	A
	Methionine	Threonine	Lysine	Arginine	G
G	Valine	Alanine	Aspartic acid	Glycine	U
	Valine	Alanine	Aspartic acid	Glycine	C
	Valine	Alanine	Glutamic acid	Glycine	A
	Valine	Alanine	Glutamic acid	Glycine	G

Figure 18.2 ▲

The genetic code. Each protein is specified by a three-letter "triplet" codon combination of adenine, guanine, cytosine, and uracil. Note how most amino acids can be specified by just the first two letters, and the third letter makes no difference—it is adaptively neutral, and all mutations at this locus are silent and nonselective. (From Donald Prothero, *Evolution: What the Fossils Say and Why It Matters*, 2nd ed. [New York: Columbia University Press, 2017])

selection. The idea that much of the genome was apparently not affected by natural selection, and thus selectively neutral, was developed at length by Motoo Kimura in his 1968 paper, "Evolutionary Rates at the Molecular Level," and an even more radical argument appeared in 1969 in a paper by J. L. King and Thomas Jukes called "Non-Darwinian Evolution." Kimura, King, and Jukes soon developed what became known as the "neutral theory of evolution."

This came as a shock to the community of evolutionary biologists at that time, and I vividly remember taking evolutionary biology courses from hard-core neo-Darwinians at Columbia University in the late 1970s who still believed in strict pan-selectionism; that is, every variation, no matter how slight, was under the control of natural selection, whether we could detect it or not. This idea goes back to Darwin himself who wrote in 1859 in *On the Origin of Species* that every variation was scrutinized (see chapter epigraph).

But at the very time I was hearing these outdated notions, the neutral theory of evolution was becoming more and more established. Surely no one could deny that the third position on a codon is nearly always silent, so any random mutation in this position would have no effect on the resulting protein and thus would be invisible to selection. At the very least, about a third of the genetic code is completely neutral and cannot be seen or affected by external selection.

The redundancy of the genome became even more apparent when a series of studies showed that most organisms have much more genetic material than they need, and most of it is simply noncoding DNA (nicknamed "junk DNA") that isn't read and has no effect on the organism. The first evidence came from a famous experiment by Richard Lewontin and Jack Hubby in 1966, which showed (by the now antique method of gel electrophoresis) that most organisms have much more variability in their DNA than they need, and that most of the DNA must therefore be unread and redundant. As molecular biology matured in the 1970s and 1980s, reading amino acid sequences became more common, and by the year 2000 the complete DNA of an organism could be read, so this redundancy became more and more apparent.

As John Sundman wrote in his 2013 article, "How I Decoded the Human Genome,"

> Kent spoke to me in nerdspeak, with geekoid locutions such as the use of "build" as a noun: "That's the most recent build of the genome. Build 31." I was used to hearing biologists talking about the elegance of DNA with what might be called reverence. By contrast Kent spoke of DNA as if it were the most convoluted, ill-documented, haphazardly maintained spaghetti code— not God's most sublime handiwork, but some hack's kludge riddled with countless generations of side effects, and "parasites on parasites."
>
> "It's a massive system to reverse-engineer," he said. "DNA is machine code. Genes are assembler, proteins are higher-level languages like C, cells are like processes . . . the analogy breaks down at the margins but offers useful insights." It was nearly impossible to tell the working code from cruft, Kent said. "That's why a lot of people say, 'The genome is junk'. " But that's what he found interesting: a high-quality programmer's code is always self-evident, but legacy assembler handed down from generation to generation of bricoleurs (I'm paraphrasing again) provides a real challenge for people who like puzzles.[1]

The most striking evidence of redundancy is the fact that the size of the genome often bears no relation to the complexity of the organism. Known as the "C-value paradox," or the "onion test," it flies in the face of the simplistic notion that more complex organisms must have larger genomes to code for all that complexity. The nickname was coined by Canadian biologist T. Ryan Gregory, and it refers to the fact that a common onion has 5 times more DNA than a human, yet it is much simpler than a human! Some salamanders have up to 35 times as much DNA as humans, and lungfish have 40 times as much DNA as humans. One species of deer has 20 percent more DNA than its close relative, and one species of puffer fish has 100 times as much DNA as another. Among plants, there is no correlation between complexity and DNA, the broad bean has 4 times more DNA than a kidney bean. Even some single-celled microbes have more DNA than humans, and simple roundworms (nematodes) and watercress have about the same amount of DNA as a human. At best, there is only a rough correlation between DNA amount and complexity (figure 18.3).

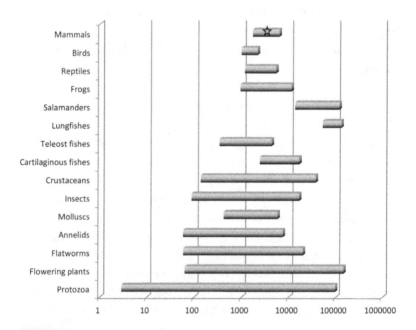

Figure 18.3 ▲

The amount of DNA in different groups of organisms. Clearly, there is no simple relationship between the amount of DNA and the complexity of the organisms. (Courtesy of T. Ryan Gregory)

Single-celled organisms often have smaller genomes than complex organisms, but there are lots of exceptions to the idea that more DNA is required for increased complexity. If an onion has 5 times as much DNA as we do, and salamanders and lungfish have 35 to 40 times as much DNA as humans, clearly most of this DNA is not coding for more structures—it is noncoding and doing nothing.

Second, it's possible to delete some of this repetitive noncoding junk sequence, and nothing happens. In 2004, an experiment deleted almost 3 percent of the mouse genome that appeared to be repetitive and noncoding, and the mice continued to reproduce with no ill effects. If this DNA were functional, how could the mice keep on reproducing without it?

So what does all this useless DNA do, if anything? Some of it may function to maintain the spacing between coding regions, or it may be used to help hold the shape of the complex folds of long DNA strands. Some of these noncoding regions include the following:

1. Introns: chunks of DNA that are initially read but then edited out during final gene splicing.
2. Pseudogenes: chunks of DNA that have lost their ability to code for proteins.
3. Repetitive DNA: in many parts of the genome, the DNA is made of the same codons repeated over and over again hundreds of times, and apparently coding for nothing.
4. Transposons: "jumping genes" that can jump from one part of the DNA to another and yet are not expressed.
5. SINEs (short interspersed nucleic elements) and LINEs (long interspersed nucleic elements): segments of DNA stuck in the middle of a coding sequence that have no function or ability to code for proteins.
6. Highly conserved noncoding nonessential DNA: consistent in the sequences of many organisms, suggesting that it is important, yet it can be removed with no effect whatsoever.

Perhaps the most intriguing of all these useless genes are endogenous retroviruses (ERVs). Most viruses work by modifying the machinery of host cells, so the host is forced to make more copies of the virus. But retroviruses (including HIV, the virus that causes AIDS and another virus that causes chicken pox/shingles) are even more insidious. They insert their own genetic code into the DNA of the host organism, and the host makes copies of the virus directly. But in many ERVs the switch to turn them on and make more virus copies has been shut off. They are basically fossil infections, old pieces of genetic code that infected our ancestors millions of years ago and have been passively copied and recopied ever since. Extinct ERVs make up as much as 5 to 8 percent of the human genome, but they are completely useless junk DNA that we carry around forever. Even more intriguing, humans share many ERVs with our ape relatives, so these infections must have occurred before humans and apes diverged more than 7 million years ago. And some ERVs are carried through most of the vertebrate family tree, indicating that these fossil infections are many millions of years old—and still they are not removed from our DNA.

Conventional pan-selectionist biologists have trouble addressing all this evidence, and instead they cling to *any* bit of biology that seems to support their belief system. In 2012, the media made a big fuss when the ENCODE project (an acronym for Encyclopedia of DNA Elements) argued that maybe 80 percent of the genome did produce some kind of protein. Naturally, many biologists jumped on this to confirm their belief that *all* of the DNA was functional. But the ENCODE project conceded that at least 20 percent of the DNA is clearly noncoding and provides no comfort to pan-selectionists, but they haven't noticed this and proclaim they have been vindicated.

It turned out that the ENCODE results were too good to be true. A study by Dan Graur and colleagues in 2013 completely demolished their work, reaffirming that indeed most of the genome (at least 90 percent of it, perhaps as much as 98 percent) is noncoding. The salient point is that the ENCODE study only managed to show that some of the genome called "junk" does indeed code for a protein. What they did *not* show is whether these random isolated proteins are part of a functional biochemical pathway, or lead to any phenotypic consequences. If a protein results from junk DNA but doesn't do anything, it's still junk.

The fact that most of the DNA in any organism is selectively neutral and apparently codes for nothing explained another discovery made in the 1960s. In 1962, the legendary scientists Linus Pauling and Émile Zuckerkandl noticed that the number of genetic differences between two related species was proportional to how long ago they had split into different lineages. The longer two species had been split into two different species, the more genetic differences they had accumulated. As another famous molecular biologist, Emanuel Margoliash, wrote in 1963:

> It appears that the number of residue differences between cytochrome c of any two species is mostly conditioned by the time elapsed since the lines of evolution leading to these two species originally diverged. If this is correct, the cytochrome c of all mammals should be equally different from the cytochrome c of all birds. Since fish diverges from the main stem of vertebrate evolution earlier than either birds or mammals, the cytochrome c of both mammals and birds should be equally different from the cytochrome c of fish. Similarly, all vertebrate cytochrome c should be equally different from the yeast protein.[2]

This remarkable discovery suggests that their DNA was constantly changing over time. The longer it had been changing, the more differences each species had accumulated. Even more surprising, it was highly consistent with the time of divergence, so most vertebrates differed by only 13 to 14 percent in molecules like cytochrome c, whereas the cytochrome c of more distantly related organisms like plants and yeast differed by 64 to 69 percent. This not only suggested that molecules could be used to reconstruct the family tree of life (see figure 7.3) but also indicated that the rate of change was roughly constant, like the ticking of a clock. This notion came to be known as the "molecular clock," and it allowed molecular biologists to estimate how long ago various lineages had branched off from their common evolutionary tree.

The idea that change slowly accumulated in the DNA over time, like the ticking of a clock, was consistent with the idea of neutralism and noncoding DNA. These changes could not slowly accumulate due to random genetic accidents and genetic drift unless they were invisible to natural selection. If most of the DNA were under the strict control of natural selection, there would not be such a tight match between divergence time and genetic distance. The evidence for neutralism from the invisible third position on the codons, the presence of so much noncoding DNA, and the fact that DNA ticks away at a constant rate while ignoring the effects of natural selection paints a consistent picture of genes: Most of the DNA is unread and is neutral or invisible to natural selection. This was a shocking idea to many biologists weaned on the pan-selectionist notions of the 1960s.

Another common dogma of early molecular evolution studies was that each gene coded for one protein, which then combined to form one distinct feature. This was known as the "one gene, one protein" dogma. But the discovery of junk DNA completely shatters this old idea. Not only do most genes code for no protein at all but some other genes code for more than one protein (known as pleiotropy). If selection maintains a certain pleiotropic gene because it codes for one very important feature, then all the other features it determines may be passively "carried along" even if they are selectively neutral or slightly harmful.

Our big toes are a case in point. Lots of ink has been spilled speculating about why humans have a big toe and how it might be selected for in our gait and stride. But this is pointless. The big toe is enlarged because it is embryonically and biochemically linked to our enlarged thumbs, which are highly adaptive. You can't have one without the other. In fact, some studies suggest that our big toes give us problems when running and walking, so there might be weak selection against them—but this is overcome by the advantages of the pleiotropic linkage with the large opposable thumb. In this case, natural selection might even allow harmful features to persist simply because they are tied to more important features with strong positive selective advantage. In addition, many features require multiple genes just to produce one feature, so the complexity is far beyond what molecular biologists could have guessed in the 1960s.

Now, more than 67 years since the structure of DNA was first discovered, we have moved away from the simplistic notions that "one gene codes for one protein" and that DNA is perfectly designed, constantly fine-tuned by natural selection, and a marvel of adaptation, or "God's handiwork." Instead, DNA and the entire genetic coding system is clumsy, inefficient, jury-rigged, and mostly composed of wasted material and unnecessary parts. It seems less well designed than some organs in the body that work efficiently. To repeat Kent's words, it is "the most convoluted, ill-documented, haphazardly maintained spaghetti code—not God's most sublime handiwork, but some hack's kludge riddled with countless generations of side effects, and 'parasites on parasites.'"

NOTES

1. John Sundman, "How I Decoded the Human Genome," *Salon*, October 21, 2003, https://www.salon.com/2003/10/21/genome_5/.

2. Emanuel Margoliash, "Primary Structure and Evolution of Cytochrome c," *Proceedings of the National Academy of Sciences of the United States of America* 50 (1963): 672–679.

FOR FURTHER READING

Carey, Nessa. *Junk DNA: A Journey Through the Dark Matter of the Genome*. New York: Columbia University Press, 2015.

Doolittle, W. Ford. "Is Junk DNA Bunk? A Critique of ENCODE." *Proceedings of the National Academy of Sciences of the United States of America* 110, no. 14 (2013): 5294–5300.

Gregory, T. Ryan, ed. *The Evolution of the Genome*. Amsterdam: Elsevier, 2005.

Graur, Dan, Yichen Zheng, Nicholas Price, Ricardo B. R. Azevedo, Rebecca A. Zufall, and Eran Elhaik. "On the Immortality of Television Sets: 'Function' in the Human Genome According to the Evolution-Free Gospel of ENCODE." *Genome Biology and Evolution* 5, no. 3 (2013): 578–590.

Hubby, J. L., and R. C. Lewontin. "A Molecular Approach to the Study of Genic Heterozygosity in Natural Populations: I. The Number of Alleles at Different Loci in *Drosophila pseudoobscura*." *Genetics* 54, no 2 (1966): 577–594.

Kimura, Motoo. *The Neutral Theory of Molecular Evolution*. Cambridge: Cambridge University Press, 1983.

——. "The Neutral Theory of Molecular Evolution." *Scientific American* 241, no. 5 (1979): 98–129.

King, Jack Lester, and Thomas H. Jukes. "Non-Darwinian Evolution." *Science* 164, no. 3881 (1969): 788–798.

Levinton, Jeffrey S. *Genetics, Paleontology, and Macroevolution*. 2nd ed. Cambridge: Cambridge University Press, 2001.

Lewontin, Richard C. *The Genetic Basis of Evolutionary Change*. New York: Columbia University Press, 1973.

Lewontin, Richard C., and J. L. Hubby. "A Molecular Approach to the Study of Genic Heterozygosity in Natural Populations: II. Amount of Variation and Degree of Heterozygosity in *Drosophila pseudoobscura*." *Genetics* 54, no. 2 (1966): 595–609.

Mindell, David P. *The Evolving World: Evolution in Everyday Life.* Cambridge, Mass.: Harvard University Press, 2006.

Nelson, P. N., P. Hooley, D. Roden, H. Davari Ejtehadi, P. Rylance, P. Warren, J. Martin, and P. G. Murray. "Human Endogenous Retroviruses: Transposable Elements with Potential?" *Clinical and Experimental Immunology* 138, no. 1 (2004): 1–9.

Ohta, Tomoko. "Near-Neutrality in Evolution of Genes and Gene Regulation." *Proceedings of the National Academy of Sciences of the United States of America* 99, no. 25 (2002): 16134–16137.

Palazzo, Alexander F., and T. Ryan Gregory. "The Case for Junk DNA." *PLoS Genetics* 10, no. 5 (2014): e1004351.

Ridley, Mark. *Evolution.* 2nd ed. Cambridge, Mass.: Blackwell, 1996.

Wesson, Robert. *Beyond Natural Selection.* Cambridge, Mass.: MIT Press, 1991.

Wills, Christopher. *The Wisdom of the Genes: New Pathways in Evolution.* New York: Basic Books, 1989.

LEGS ON THEIR HEADS

You have loaded yourself with an unnecessary difficulty in adopting Natura non facit saltum ("Nature does not make leaps") so unreservedly.

—THOMAS HENRY HUXLEY, IN AN 1859 LETTER TO CHARLES DARWIN

As discussed in chapter 15, the idea that acquired characteristics could be passed from parent to child was widespread in the days before modern genetics. It has been mislabeled "Lamarckism," but even Darwin believed it. The appeal of this idea is obvious. It would allow organisms to adapt rapidly, in just one generation, in direct response to environmental demands. Classic Darwinian selection, in contrast, is very slow and wasteful. Many offspring are born but only a few survive with the favorable variations, and many generations are required for a whole population to become established as a new species.

Darwin's ideas transformed biology, and geneticists began to more rigorously test the idea of acquired inheritance. German biologist August Weismann ran a series of experiments in the 1880s that seemed to discredit acquired inheritance once and for all. Weismann cut off the tails of 20 generations of mice, but each succeeding generation of mice was born with normal tails, not shorter tails in response to this extreme environmental pressure. Weismann concluded that changes occurring in our phenotype ("soma" in Weismann's terminology) cannot ever get back into the genotype (what Weismann called the "germ line"). In other words, the flow of information is strictly one way. Changes in the genotype dictate changes in the phenotype, but not the other way around. This came to be known as the "central dogma" of molecular genetics. Later, James Watson, codiscoverer of the structure of DNA, redefined the central dogma to mean the one-way path from DNA to RNA to protein.

Over the years, various scientists have proposed ideas that appear to violate the central dogma. In the 1950s, embryologist Conrad Waddington subjected larval fruit flies to heat shock and produced a mutation in which the wings lacked a cross vein. This procedure was carried on generation after generation, and after 14 generations crossveinless flies appeared without administering heat shock. Had the environmental stress somehow changed the genotype directly rather than through selection? Waddington called this phenomenon "genetic assimilation," and neo-Darwinists continue to argue over how to explain it without neo-Lamarckism.

More recently, immunologists have conducted experiments that seem to show acquired inheritance. Whenever an organism is exposed to a disease, its immune system develops antibodies that kill the foreign infection. This immunity is acquired during one's lifetime and should not be able to work its way back into the genome. However, experiments have shown that laboratory mice could pass on their immunity to their offspring. Although neo-Darwinists are still arguing that this can be explained by non-Lamarckian means, it raises serious questions about the inviolability of the germ line.

In molecular biology, more and more examples have been documented in which genes were changed after the organism was born. Barbara McClintock won the Nobel Prize for her discovery of "jumping genes" that move from one spot on the DNA strand to another, changing the gene code. Other experiments have shown that external DNA can be incorporated into a cell and possibly into the host DNA. In one case, different bacteria appeared to exchange bits of genetic material, a switch that allowed them to all evolve a new mutation rapidly. One group of viruses, retroviruses (such as the HIV virus that causes AIDS and also the virus that causes chickenpox/shingles), copy their own genetic information into the DNA of the host. Could this be the mechanism that allows environmental changes to be translated directly into the genetic code?

Although hard-core neo-Darwinians are still debating and disputing the implications of these studies, it is now clear that the genome is far more complicated and flexible than the original static entity visualized by

Weismann. As molecular biology and immunology find more and more exceptions to the central dogma, the once disreputable idea that organisms can respond to environmental stresses by changing their genomes directly may no longer be so outrageous.

The first breakthrough in discovering how gene regulation works was published in 1961 by Jacques Monod, Jean-Pierre Changeux, and François Jacob. They were working with the familiar bacterium *Escherischia coli* (the coliform bacteria also known as *E. coli*, which lives by the millions in our gut flora). They found that it had a cluster of genes that worked in a feedback loop, and switched on or off a particular gene, the lac operon, that produced the enzyme lactase, used to break down milk sugars (lactose). To their surprise, they found that the switch that turns the lac operon on or off is an external stimulus in the environment. If there is too much lactose in the surroundings of the bacterium, the switch is turned on, and the bacterium rapidly produces lots of lactase to digest the excess sugars. If the lactose concentration drops, the switch is turned off and the bacterium no longer invests energy in breaking down lactose. In other words, an external substance triggers the ability of the genome to produce certain things. Clearly, the genome does not ignore the environment and can be affected by external stimuli.

The most important breakthroughs in the genetics of development came from Edward B. Lewis in 1978 and his discovery of homeotic genes, which can make huge and abrupt transformations by producing normal anatomical parts in odd places. For example, normal flies have two sets of wings and two balancing organs, called halteres, where the second pair of wings would be. A homeotic mutation produces the "antennipedia" condition, in which a normal fly grows a leg out of its head in place of an antenna (figure 19.1A). Another homeotic mutation changes these halteres back into the ancestral wings, producing four-winged flies (figure 19.1B). More recently, it has been discovered that a series of homeotic genes known as the Hox complex control the basic segmentation of the body, not only of arthropods but also of vertebrates. Clearly, homeotic genes are fundamental to the body plans of almost all animals, and a small change in the homeotic genes can have huge effects on the phenotype, producing new body plans, extra limbs, or extra segments in a single generation. This directly contradicts the old neo-Darwinian assertion that novel features arose only by gradual selection over many generations.

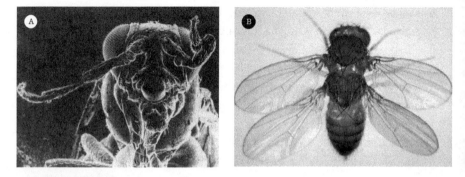

Figure 19.1 ▲

Homeotic mutants show that big developmental changes can result from small genetic mutations, giving rise to dramatic differences in body plan. (*A*) The antennipedia mutation, where a leg grows instead of an antenna. (*B*) The bithorax mutant fly, which has a second pair of wings instead of the halteres normally found behind the front pair of wings. ([*A*] Courtesy of F. R. Turner; [*B*] courtesy of Walter Gehring and G. Backhaus)

From these early discoveries, molecular biologists have identified most of the Hox genes in a number of organisms and found that nearly all animals (including flies, mice, and humans) use a very similar set of Hox genes, with slight variations and additions. Each Hox gene is responsible for the development of part of the organism and all of its normal organ systems (figure 19.2). Some homeotic genes work at the fundamental level and are found in all living things and control similar parts of development. For example, many of the Hox genes that control the basic parts of animal development are also found in fungi, yeasts, and plants, but controlling different structures.

Small changes in the Hox genes can put different appendages on a segment of a fly (such as the leg where the antenna would go, or the wing where a haltere belongs), or even multiply the number of segments. The key Hox gene in this case is called "Distal-less," and it controls the development of the limbs in nearly all animals: insect appendages, fish fins, chicken limbs, the bristles of marine annelid worms, the ampullae and siphons of the tunicates or "sea squirts," and the tube feet of sea urchins. Thus it must be an ancient controlling gene, dating back over 600 million years ago when nearly all the major phyla of animals diverged from a common ancestor. A tiny change in Hox genes can make a big evolutionary difference. In the arthropods (the "jointed legged" animals, such as insects, spiders,

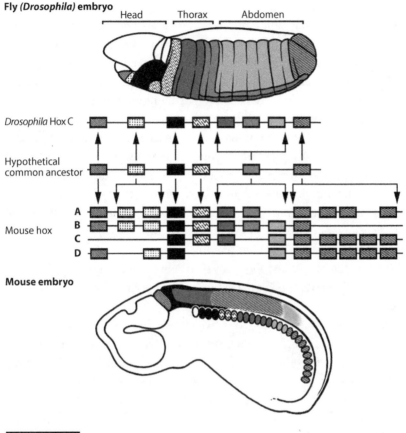

Fly (Drosophila) embryo

Figure 19.2 ▲
Map of the locus of action of the Hox genes in the fly and in the mouse. Note that the basic Hox genes are similar in almost all bilaterally symmetrical animals, so the system goes back to the very origin of complex animals. Small changes in any of these Hox genes make big differences in body plans. (Drawing by Carl Buell)

scorpions, and crustaceans), for example, a small change in the Hox genes can multiply the number of segments or reduce them and switch one appendage (for example, a leg) on each segment with another (for example, a crab claw or an antenna or mouth parts). Arthropods are a classic example of this modular development with interchangeable parts.

With a small change in Hox genes, whole new body plans can evolve quickly to exploit new resources. A good example is the modularity of the skeleton of arthropods. As we saw with fruit flies, Hox genes dictate whether they will have the primitive condition of two pairs of wings (as in primitive insects like dragonflies) and whether an appendage develops into an antenna, a leg, a mouthpart, or some other part. The fossil record confirms this idea that simply switching Hox genes on or off allow abrupt changes not only in appendages and wings but even in the number of body segments. For example, a number of fossils show that many primitive insects originally had more than two pairs of wings, which suggests the reduction to two pairs in dragonflies and one pair in many other insects was an abrupt homeotic change. Another study in 2009 demonstrated how whole new groups of millipedes had arisen by saltational evolution, in which they added a bunch of new segments all at once. A 2002 study put a shrimp Ubx Hox gene into an insect larva and showed how this gene was responsible for suppressing the development of limbs in insects (which have 6 legs, compared to the 10 in most crustaceans). Other scientists have manipulating Hox genes to show how you can get just about any type or number of appendages on each segment of an arthropod or make radical changes in body plan with a simple gene change.

Another key Hox gene is *pax-6*, which switches on the development of eyes in nearly all animals. Earlier studies of different types of photoreceptors and eyes in the animal kingdom showed that this had been done many different ways, and Ernst Mayr and other biologists argued that some form of eye had evolved independently at least 50 times (see chapter 20). After all, an insect or trilobite eye is made of hundreds of tiny lenses all clustered into a spherical bundle, whereas the eye of a vertebrate is a fluid-filled globe with a lens at one end, and the eye of an octopus has a similar shape but is very different in detail. But work by Walter Gehring and his colleagues since 1994 revealed that all of these very different kinds of photoreceptors and eyes were controlled by the *pax-6* gene, even though they had evolved into utterly different ways of sensing and processing light information.

All of these ideas are part of the exciting new research field known as evolutionary development (nicknamed "evo-devo"), and it is now the hottest topic in evolution. We have gone from neo-Darwinian insistence on every gene gradually changing to make a new species to realizing that only a few key regulatory genes need to change to make a big difference, often in a single generation. This circumvents many of the earlier problems with ideas about macroevolution. It is possible that the processes that build new body plans and allow organisms to develop new ecologies are not the product of small-scale microevolutional changes extrapolated over time. Some evolutionists still see evo-devo as just an extension of the neo-Darwinian synthesis, but others argue that it is an entirely different type of process from that envisioned in the 1950s.

Either way, the idea that all multicellular organisms on the planet—whether animals, plants, or fungi—share the same common genetic tool kit tells us a lot about how closely related we are to all the rest of life. Those few genetic switches found in our tool kit can produce the eyes, the limbs, and other completely different structures in everything from fruit flies to mice to humans. The discovery that a few genetic switches can produce radical changes in body structure in a short time allows us to look at the possibility of macroevolutionary change through changes in the environment, and how key genes for regulation can be affected by environmental conditions. This was something not even Lamarck could have imagined, let alone Charles Darwin!

FOR FURTHER READING

Carroll, Sean. *Endless Forms Most Beautiful: The New Science of Evo Devo*. New York: Norton, 2005.

——. *The Making of the Fittest: DNA and the Ultimate Forensic Record of Evolution*. New York: Norton, 2006.

Gregory, T. Ryan, ed. *The Evolution of the Genome*. Amsterdam: Elsevier, 2005.

Held, Lewis I., Jr. *Deep Homology? Uncanny Similarities of Humans and Flies Uncovered by Evo-Devo*. Cambridge: Cambridge University Press, 2017.

——. *How the Snake Lost Its Legs: Curious Tales from the Frontiers of Evo-Devo*. Cambridge: Cambridge University Press, 2014.

——. *Quirks of Human Anatomy: An Evo-Devo Look at the Human Body*. Cambridge: Cambridge University Press, 2009.

Levinton, Jeffrey. *Genetics, Paleontology, and Macroevolution*. 2nd ed. Cambridge: Cambridge University Press, 2001.

Ridley, Mark. *Evolution*. 2nd ed. Cambridge, Mass.: Blackwell, 1996.

Wills, Christopher. 1989. *The Wisdom of the Genes: New Pathways in Evolution*. New York: Basic Books, 1989.

THE EYES HAVE IT

To suppose that the eye with all its inimitable contrivances for adjusting the focus to different distances, for admitting different amounts of light, and for the correction of spherical and chromatic aberration, could have been formed by natural selection, seems, I freely confess, absurd in the highest degree. When it was first said that the sun stood still and the world turned round, the common sense of mankind declared the doctrine false; but the old saying of Vox populi, vox Dei, as every philosopher knows, cannot be trusted in science. Reason tells me, that if numerous gradations from a simple and imperfect eye to one complex and perfect can be shown to exist, each grade being useful to its possessor, as is certainly the case; if further, the eye ever varies and the variations be inherited, as is likewise certainly the case; and if such variations should be useful to any animal under changing conditions of life, then the difficulty of believing that a perfect and complex eye could be formed by natural selection, though insuperable by our imagination, should not be considered as subversive of the theory. How a nerve comes to be sensitive to light, hardly concerns us more than how life itself originated; but I may remark that, as some of the lowest organisms, in which nerves cannot be detected, are capable of perceiving light, it does not seem impossible that certain sensitive elements in their sarcode should become aggregated and developed into nerves, endowed with this special sensibility.

—CHARLES DARWIN, *ON THE ORIGIN OF SPECIES* (1859)

In *On the Origin of Species*, Darwin addressed one of the biggest objections to his ideas, what he called "organs of extreme perfection." He knew that examples of extraordinary design and coordination of anatomical structures were the essence of the old "natural theology" school of thought, and he took pains to emphasize examples of poor or shoddy design in chapter 9 in his book. But he could not avoid talking about beautiful design forever.

So he tackled it head on in chapter 6, especially organs like the eye. How could such a perfect structure be produced by random natural selection? On the face of it, Darwin "freely confessed" that it seemed "absurd in the highest possible degree."

Indeed, Darwin's critics, such as St. George Jackson Mivart, attacked him on this issue when his ideas were first published. The epigraph at the beginning of this chapter appeared in later editions of his book as a direct answer to Mivart. Science deniers, in particular, are fond of quoting just the first sentence of the epigraph, claiming that Darwin had admitted that the idea of evolution was "absurd in the highest possible degree." They dishonestly leave out the entire rest of the quote because, of course, the first sentence is setting up the initial dilemma, and the entire rest of the quotation shows that Darwin *does* have an answer to how the eye could have evolved. I've caught them doing this in debate, and thanks to my smartphone, I can pull up this entire quotation and read it back to them. Either they completely fail to understand what Darwin said and keep insisting that only the first sentence matters, or they are silenced on the issue and change the subject.

In 1859, Darwin didn't have much evidence available to support his case. He could point out that simple amoeba-like organisms ("sarcodes" in Darwin's time) did have light-sensing cells in their tissues that helped them navigate to or from the light, and he provided three pages of examples of organisms with different kinds of light perception. But very little work had been done on photoreception in other groups of animals in Darwin's time. Fortunately, in the more than 160 years since Darwin's book was published, an enormous amount of work on light sensing in the animal kingdom has been done, and now we have an extraordinary sequence of different kinds of photoreceptors and eyes that show how sight could indeed evolve. (If you are interested, I recommend Ivan Schwab's beautifully illustrated 2012 book, *Evolution's Witness: How Eyes Evolved*, which has hundreds of illustrations and color photographs showing nearly every kind of eye known.)

Or, in Darwin's words:

Reason tells me, that if numerous gradations from a simple and imperfect eye to one complex and perfect can be shown to exist, each grade being useful to its possessor, as is certainly the case; if further, the eye ever varies and the variations be inherited, as is likewise certainly the case; and if such variations should be useful to any animal under changing conditions of life, then the difficulty of believing that a perfect and complex eye could be formed by natural selection, though insuperable by our imagination, should not be considered as subversive of the theory.

The first stage of this transition, from blind creatures to those that can perceive light, occurs in many single-celled organisms often called "Protista" or "protozoans," some of which were called "sarcodes" in Darwin's time. Many of them have light-sensitive proteins, especially those that also carry chloroplasts and are photosynthetic, such as *Euglena*, so they must seek out light. All that an organism needs for photoreception is the light-sensitive protein opsin that surrounds pigmented cells called chromophores, which can distinguish colors (figure 20.1A). Such simple "eyespots," a cluster of light-sensitive cells, are thought to have evolved independently, possibly 40 to 65 times, in many different unrelated lineages of animals. However, these eyespots are controlled by the same homeobox genes, especially the *pax-6* Hox gene (see chapter 19), so the genetic blueprint for making some kind of simple eye is universal in the animal kingdom and even in the single-celled protistans. It occurs not only in single-celled creatures but also in some sea jellies such as *Cladonema*, which has photoreceptors but no brain. The electrical messages of their eyes are transmitted directly to the muscles to stimulate movement with respect to the light source. In almost all higher organisms, the chromophores in the photoreceptor absorb photons of light, which are turned into an electrical signal that moves through the nervous system to whatever central ganglion or brain is found in the organism.

The next stage in eye evolution from the eyespot is a simple bowl- or cup-shaped light receptor (figure 20.1B). Eyespots cannot determine the direction from which light comes, but a cup-shaped receptor allows an animal to detect the angle and direction of the light source because it hits only certain parts of the cup-shaped surface over and over again. This is the kind of eye found in flatworms or planarians, which are among the simplest animals with bilateral symmetry and a head and a tail (something not found in more primitive radial creatures called Cnidaria, like sea jellies or sea anemones, or in sponges). The cup-shaped eye is also found in most snails, both living and presumably among the extinct ones, which would have been one of the first creatures in the Early Cambrian with the ability to detect light. Most of

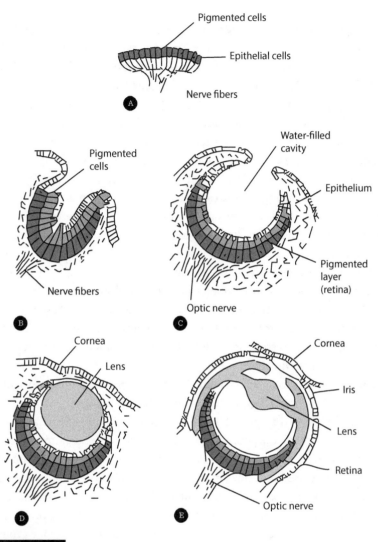

Pigmented cells

Epithelial cells

Nerve fibers

A

Pigmented cells

Water-filled cavity

Epithelium

Nerve fibers

Pigmented layer (retina)

Optic nerve

B

C

Cornea

Lens

Cornea

Iris

Lens

Retina

Optic nerve

D

E

Figure 20.1 ▲

The stages of evolution in photoreceptors. (*A*) The simplest photoreceptors are patches of light-sensitive pigmented cells in the outer layer of tissue in many simple organisms. (*B*) A slightly more advanced photoreceptor is a bowl- or cup-shaped pocket in the dermis, with photoreceptors in the bottom that help discriminate the direction from which the light and dark are coming. (*C*) A more advanced type of eye is a deep round pocket with a small opening, creating a pinhole camera effect that projects a low-resolution image of the world onto the retina. (*D*) Even more advanced is a spherical eye enclosed by a clear corneal membrane, which has many advantages over other types of eyes. (*E*) The most advanced eyes are enclosed eyeballs with not only a cornea but also a lens and iris mechanism, so the eye can form sharp images. This type of eye evolved independently in different ways in the squids and octopus and in the vertebrates. (Courtesy of Wikimedia Commons)

these organisms use the ability to detect the direction of light to find shade and shelter or to detect the shadow of a predator swimming overhead.

In some animals, the simple cup-shaped eye becomes deeper and has a much greater density of photoreceptive cells, so the eye is capable of processing much more information about the environment that just the general direction of the light source. Eventually, the cup gets deeper and deeper until the opening is restricted in size, creating a simple pinhole camera effect (figure 20.1C). A pinhole opening, combined with a deep cup-shaped retina with a dense layer of photoreceptive cells, allows not only directional sensing but also some shape-sensing and dim imaging. It reduces distortion and scatter by allowing only a thin beam of light into the eye. But such a pinhole-type light sensor does not have a cornea or a lens, so the eye is only capable of very poor resolution and dim imaging. However, it is a tremendous improvement over the bowl-shaped eye of a flatworm or a snail. Among molluscs, this pinhole-shaped eye is found in the chambered nautilus. Presumably, this advanced type of eye was also present in the Late Cambrian, when the earliest relatives of nautiloids first evolved, and it gave these enormous straight-shelled nautiloids the ability to find prey and become the earth's first large predators.

The next step is to cover the pinhole opening with a layer of transparent cells, forming a cornea (figure 20.1D). By sealing off the inner chamber, the eye could develop a transparent fluid filling (called a "humor" in anatomical terms), which gives it the potential for color filtering or for creating a higher refractive index, blocking UV radiation, as well as giving it the potential to operate in air as well as it does in water. It also prevents contamination of the inside of the eye, or invasion by infection or parasitic infestation. The presence of a cornea also increases the refractive power of the eye. Simple eyes like this are found in many parts of the animal kingdom. One group, the "velvet worms" or Onychophora, have a simple spherical fluid-filled eye with a cornea. When they molt, their entire outer skin including the corneal covering is shed with the rest of the exoskeleton. It will have one or two cuticular layers of cornea, depending on how recently it has molted.

The next stage of eye evolution is to develop a layer of cells made of a protein called lens crystallin on the inside of the cornea to form a lens (figure 20.1E). There are two different types, alpha-lens crystallin and beta-gamma lens crytallin, both of which are proteins originally developed for other functions that were co-opted to become part of the lens of the eye.

These proteins are unique in having high transparency, the ability to pack tightly, and the ability to last through the organism's entire life span.

The cells of the eye are a living tissue in an embryo, but once the organism is born, the opaque cellular machinery that produced the lens proteins must be removed and the lens becomes a matrix of dead cells packed with crystallins. Maintaining a layer of transparent cells is a problem for most animal eyes. In some organisms, such as trilobites (figure 20.2), the lens crystallins were replaced with the mineral calcite. Sometimes these fossilized eyes are so well preserved that paleontologists have been able to look through them and see and photograph what the trilobite originally could see. In some more advanced trilobites, the lens is a doublet with a concave lower portion and a convex upper portion, which corrects for spherical aberration in thick lenses. The trilobites invented this solution to vision problems some 400 million years ago, and humans only rediscovered it in the 1700s.

The addition of a lens is a huge improvement. In the simple eye with only a cornea, the opening yields a broad patch of light over a large area of the retina. But the addition of a lens focuses the incoming visible radiation on a small area of the retina, allowing it to sense even dim light and to form a much better image in bright light. The lens proteins also have a high refractive index, allowing them to focus light more acutely.

From this development, the arrangement and number of lenses in the eyes have varied widely across the animal kingdom. Most of the jointed legged animals, or arthropods (insects, spiders, scorpions, crustaceans, millipedes, centipedes, most trilobites, and their kin), have many small lenses packed into a compound eye. Each tiny lens senses a small difference in light or dark, and when combined together the brain of the animal can put together a complex mosaic of light and dark patches that it interprets as an image. Many of these compound eyes have hundreds of lenses and bulge outward almost into a spherical shape, so the animal can see almost all sides around it, warning it of predators coming from any direction.

In other phyla, a different sequence of evolutionary stages can be traced across the members of the group. The molluscs, for example, have the widest range of eye types found in any phylum. Some of the most primitive molluscs are blind, or have small patches of light-sensing cells scattered all over their bodies to detect the dark shadow of an oncoming predator. The next stage is the primitive flat eye, or eye spot, where a bundle

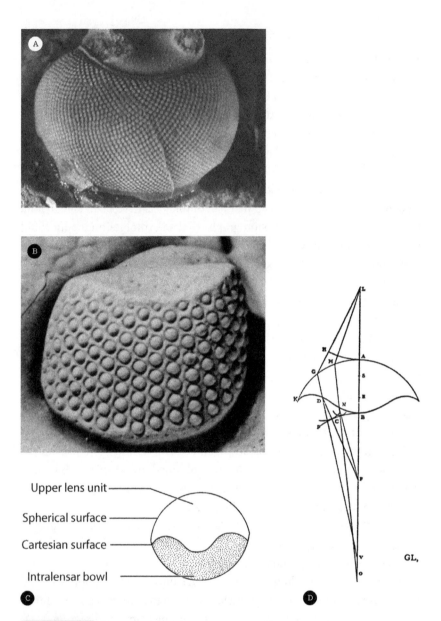

Upper lens unit
Spherical surface
Cartesian surface
Intralensar bowl

GL,

Figure 20.2 ▲

Anatomy of the eyes of trilobites. (*A*) Holochroal eye composed of closely packed lenses.
(*B*) Schizochroal eye with lenses separated by solid cuticle. (*C*) Cross section of the lens of
a schizochroal eye, showing the division into two discrete units. (*D*) Huygens's 1690 diagram
of the ray path optics of lenses, illustrating how they correct for spherical aberration. The
upper lens unit of trilobite eyes solved this problem 40 million years before humans did.
([*A*, *B*] From Riccardo Levi-Setti, *Trilobites* [1758; repr., Chicago: University of Chicago Press,
1993]; [*C*] from Richard S. Boardman, Alan H. Cheetham, and A. J. Rowell, *Fossil invertebrates* [Palo Alto, Calif.: Blackwell, 1987]; [*D*] Public domain)

of light-sensing cells are clustered together. They allow the organism not only to sense dark and light but to tell from where the light is coming. This kind of eye is found today not only in sea jellies but also in a handful of very primitive molluscs.

The next step is the cup-shaped eye with a pigmented light-sensing layer at the bottom. These are widespread among molluscs that need to detect an oncoming predator, and they are found in limpets, chitons, and certain types of clams. This was followed by the pinhole-camera style of eye, which gives a focused, low-light picture on the retina. These are found in the nautilus and also in abalones and their relatives. The more advanced eye has a cornea (here, a translucent epithelium) but no permanent lens to speak of, although some molluscs use a vesicle or bubble of fluid to serve as a lens. These are the kinds of eyes found in predatory marine snails and most land snails, and, surprisingly, in scallops, which have numerous small eyes around the edge of their mantle that are able to detect the approach of a predator.

The most advanced molluscan eyes have fully sealed spherical eyeballs filled with fluid and a crystalline lens, an iris to regulate incoming light, and several other features that look remarkably like a vertebrate eyeball. Superficially, they may look the same, but in detail they are quite different. In vertebrate eyes, the ciliary muscles around the lens are used to squeeze the flexible lens or relax it, so it changes shape and can focus the light better (figure 20.3). But molluscs use the same muscles to move the lens forward and back, shifting the focal plane until there is good focus, without changing the shape of the lens. Even more striking is the configuration of the light-sensing cells in the retina. In molluscs, like the octopus and squid and cuttlefish, these cells are on the top layer of the retina, so they directly face the light source. In contrast, in a vertebrate eye the light-sensing cells point backward, away from the light source, and are covered by the network of nerves and blood vessels that sustain them. This somewhat distorts the image coming into the light sensors, giving us inferior vision compared to an octopus. In addition, this clumsy design results in the nerve bundle that connects it to the brain coming through the middle of the retina, creating a blind spot. The octopus eye has no blind spot and once again is superior to the vertebrate eye in this regard. Octopus, squid, and cuttlefish put their excellent eyes to good use. Not only do they see their prey well to capture it, and flee from predators (they lack an external

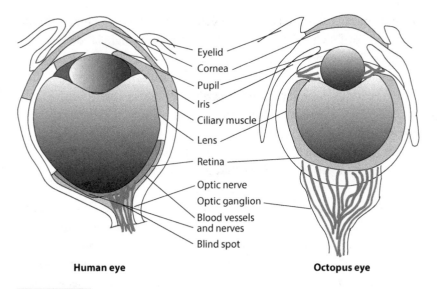

Eyelid
Cornea
Pupil
Iris
Ciliary muscle
Lens
Retina
Optic nerve
Optic ganglion
Blood vessels and nerves
Blind spot

Human eye   **Octopus eye**

Figure 20.3 ▲

Comparison of the advanced eyes of an octopus and a vertebrate. Both eyeballs have a fluid-filled center, a cornea, lens, and iris, and the ability to see sharp complex images, often with color vision as well. But the vertebrate eye has a major design flaw: the photoreceptors (rods and cones) are beneath the layer of nerves and blood vessels supporting them and are pointed the wrong way, which limits the vision of this kind of eye. It also has an optic nerve that must connect to these layers of tissue on top of the retina, then exit through a hole in the retina, creating a blind spot. In contrast, the octopus eye has neither of these flaws: the photoreceptors point toward the light source and are wired from beneath, connecting directly to the optic nerve, so there is no blind spot. No matter how well designed our eyes seem, they are jury-rigged and poorly designed when compared to the eyes of an octopus. (Redrawn from several sources by E. Prothero)

shell like the nautilus and most other mollusks) but they also perceive their background clearly and can change their skin patterns to camouflage themselves against any setting. In addition, these amazing creatures communicate by flashing an incredible variety of colors and patterns across their skin in fractions of a second. If you have never watched these mesmerizing displays, I highly recommend looking at a few videos of their changing coloration on the internet.

Vertebrates mostly have eyeballs similar to our own and that we think of as "normal" for animal eyes. But the most primitive vertebrates, such as the slimy jawless hagfish, has only a simple cup-shaped eye. It lives in deep dark murky waters and navigates and feeds by taste and smell. Our primitive nonvertebrate relatives, such as the tunicates (sea squirts) and amphioxus or lancelet, have a simple eye spot, at least as larvae. And fossils of the earliest jawless fish from the Cambrian of China seem to have a simple eye spot as well. But starting with lampreys, which have an eye with a lens but no real fluid-filled eyeball, and then sharks, which have an eyeball much like our own, nearly all vertebrates have a more complex eye with lens, iris, and all the other features we associate with our vision.

Thus Darwin's challenge has been met. He suggested that there must have been many steps in making a complex eye from a simpler eye, and he mentioned the few examples that were known in his time. We have seen that there are in fact numerous intermediate steps from a simple eye spot through complex eyes with lenses and other features, and several groups, such as molluscs and vertebrates, have independently gone through all of these steps to evolve the complex eyeball that we have. He suggested that "numerous gradations from a simple and imperfect eye to one complex and perfect can be shown to exist," and science has now documented all those gradations, not only in our more primitive living relatives but in some cases in the fossil record as well. Imagining how the eye evolved is no longer "absurd in the highest degree."

FOR FURTHER READING

Glaeser, Georg, and Hannes F. Paulus. *The Evolution of the Eye*. Berlin: Springer, 2015.

Lamb, Trevor D. "Evolution of the Eye." *Scientific American* 305, no. 1 (2011): 64–69.

Land, Michael F. *Eyes to See: The Astonishing Variety of Vision in Nature*. Oxford: Oxford University Press, 2019.

Land, Michael F., and Dan-Eric Nilsson. *Animal Eyes*. 2nd ed. Oxford: Oxford University Press, 2012.

Oakley, Todd H., and Daniel I. Speiser. "How Complexity Originates: The Evolution of Animal Eyes." *Annual Review of Ecology, Evolution, and Systematics* 46 (2015): 237–260.

Parker, Steven. *Color and Vision: The Evolution of Eyes and Perception*. New York: Firefly Books, 2016.

Schwab, Ivan. *Evolution's Witness: How Eyes Evolved*. Oxford: Oxford University Press, 2012.

PART V

HUMANS AND EVOLUTION

A TINKERER, NOT AN ENGINEER

Anyone who thinks that our body is a marvel of mechanical engineering should get out a phone book and scan the listings of orthodontists, orthopedists, optometrists, and chiropractors—to name a few specialties. Those doctors are making good livings treating our sundry flaws. Face it: our body has many features that could work much better. The sad fact is that evolution is no engineer. It's just a tinkerer.

—LEWIS I. HELD JR., *QUIRKS OF HUMAN ANATOMY: AN EVO-DEVO LOOK AT THE HUMAN BODY* (2009)

In the 1600s and 1700s, naturalists and theologians (often the same person) held up the human body as an example of perfect design and engineering. After all, doesn't Genesis 9:6 say that "God made man in His own image"? Therefore, the human body must be perfect or at least as good as could be designed. This extreme view was often pushed by the "philosophical optimism" school of thought articulated by Gottfried Wilhelm Leibniz and other thinkers in the 1670s and 1680s. It was also brutally satirized by Voltaire in his 1759 novel *Candide*, in which the Leibnizian philosophy is voiced by the character of Professor Pangloss, who states that this is the "best of all possible worlds" (despite all of its apparent problems, such as death, disease, and natural disasters). Pangloss even goes so far as to spout absurdities such as "Our legs were made so we could wear breeches" or "It is demonstrable that things cannot be otherwise than as they are; for as all things have been created for some end, they must necessarily be created for the best end. Observe, for instance, the nose is formed for spectacles, therefore we wear spectacles."

The story then puts Candide, and sometimes Pangloss, through all the imaginable horrors of life, from losing his family and wealth and his ancestral home, to nearly being killed in war, to being shipwrecked and barely surviving, to arrive in Lisbon harbor only to experience the horrific 1755 Lisbon earthquake, tsunami, and fires. Candide and Pangloss are tortured and nearly executed by the Inquisition when they are blamed for the Lisbon earthquake. They escape eventually, only to become slaves in the New World. Voltaire's brilliant satire so completely discredited the school of philosophical optimism that one would think it would never return. But the power of religious dogma is strong, and the same basic idea was present in the school of natural theology popular in the 1700s (see chapter 9). It was a less absurd approach than that of Leibniz, only inferring a Divine Designer from the intricate design of nature. Nonetheless it promoted ideas similar to those of Leibniz.

The basic idea of natural theology was completely debunked by the Scottish philosopher David Hume in his 1779 book, *Dialogues Concerning Natural Religion*. But the optimistic idea that nature was evidence of God's handiwork and his divine design was still apparent in 1802 in Rev. William Paley's book, *Natural Theology*. Darwin knew Paley's book almost by heart when he was a student. But by the time Darwin wrote *On the Origin of Species* in 1859, he had seen so many examples of poor design, or of jury-rigged and suboptimal structures, in nature that he could no longer accept Paley's naïve view.

Darwin discussed the problem of suboptimality and rudimentary or vestigial organs in his 1859 book, but throughout that work he fastidiously avoided discussing humans as examples of evolution. When he wrote *The Descent of Man* in 1871, however, the issue was no longer avoidable, and he discussed the problem of vestigial and poorly designed organs in humans at length in that book:

> Not one of the higher animals can be named which does not bear some part in a rudimentary condition; and man forms no exception to the rule. Rudimentary organs must be distinguished from those that are nascent; though in some cases the distinction is not easy. The former are either absolutely useless, such as the mammæ of male quadrupeds, or the incisor teeth of ruminants which never cut through the gums; or they are of such slight service to their present possessors, that we can hardly suppose that they were

developed under the conditions which now exist. Organs in this latter state are not strictly rudimentary, but they are tending in this direction. Nascent organs, on the other hand, though not fully developed, are of high service to their possessors, and are capable of further development. Rudimentary organs are eminently variable; and this is partly intelligible, as they are useless, or nearly useless, and consequently are no longer subjected to natural selection. They often become wholly suppressed. When this occurs, they are nevertheless liable to occasional reappearance through reversion—a circumstance well worthy of attention.

As Darwin pointed out, neither humans nor any other organisms were perfectly designed; rather, they are composed of organs and tissues that work just well enough for the organism to survive and reproduce. In addition, because the purpose of natural selection is to pass on genes to the next generation, any features that develop or change after successful reproduction don't matter. For a variety of insects and other invertebrates and for fish like salmon, successfully mating and laying their eggs is all that is needed; then the adults quickly die off. In other cases, older nonreproductive adults often become injured and die or are killed by predators, making the rest of the group younger and more vigorous. Among the exceptions that prove the rule are elephants and other long-lived animals with strong societal bonds. The oldest individuals are matriarchs; they can no longer reproduce, but their strength, experience, and wisdom ensure the survival of the younger breeding females in their herd.

Anthropologists have pointed out that human societies have had elements of both of these features over time. Until recently, individuals in most hunter-gatherer groups or in simple agricultural civilizations seldom lived past the age of 30 or 40, and by that time they had either succeeded in raising children—or they hadn't. Cancer, heart disease, strokes, dementia, and other infirmities that now plague our elderly populations were not a concern in earlier societies because humans rarely lived long enough to experience these problems. Older individuals could be a burden on the tribe, and in some groups they were left to die if conditions became too harsh and the entire group was in danger of dying from starvation or some environmental stress. In other human societies, older individuals were fed and sheltered even though they might slow the tribe or require extra care. These elders retained a store of experience and tribal wisdom that the younger reproductive individuals valued. Nature's ruthless calculus favoring those who ensure the reproductive fitness of the total group also found exceptions that prove the rule in human societies. Most human societies today value the lives of all individuals.

To overcome the tendency of nature to favor only those of reproductive age and the overall fittest organisms, human social groups accept these less-fit individuals and often compensate for their weaknesses and infirmities. Those of us who are not fit enough to run with a hunter-gatherer tribe or are unable to kill game because we have bad eyesight survive today because culture overrules nature. Devices created through the inventiveness of humankind, such as glasses, hearing aids, prosthetics, and medicines, enable us to survive a lot longer than nature would normally have permitted.

Nevertheless, the jury-rigged nature of our bodies remains true. As long as these features are "just good enough" for an organism to survive (or at least don't directly harm us), they will persist. In many cases, these suboptimal features are present because our relatives had them, or our ancestors had them, and they are hard-wired in our genome even though they are no longer functional. In most cases, it's too costly to rewire our genes to get rid of these useless features. As Lewis Held Jr. put it in his *Quirks of Human Anatomy*, we still cope with

> suboptimal functioning of a structure due to (1) its being used in a novel context (e.g., a bipedal vs. quadrupedal stance) and (2) the inability of evolution to fix it (e.g., due to prohibitive demands of genomic rewiring). Tantamount to a species getting stuck on a low peak in a rugged adaptive landscape. (33)

Let's run down some of the long list of poor designs and vestigial features of humans, just to remind us of our humble origins. Many of these features were configured in a certain way in our ancestors, and this roundabout wiring and clumsy, inelegant design has been maintained even though it does not function as well as it should. One of these features is the vertebrate eye (see chapter 20). Our eyes are wired backward, with the photoreceptors in the retina pointed away from the light source, and the network of blood vessels and nerves lies on top of them in the retina, which makes our vision less acute than it could be. This configuration also necessitates having an opening for the optic nerve, which creates a blind spot in our retina. If humans were divinely designed, surely our eyes would be like those of the octopus, in which the photoreceptors point the right way and are in the top layer of the retina with nothing obstructing them, nor any blind spot (see figure 20.3).

Another famous example is the descent of the testes in humans. In cold-blooded animals, the testes are embedded in the body to keep their temperature as constant as possible. But having the testes in the abdominal cavity in warm-blooded animals often makes them too hot and can inhibit the healthy development of sperm. Some animals have an interior cooling system to prevent this, but in humans and other primates the testes emerge from the internal body cavity into a sac called the scrotum and hang loose beneath the body. This does help keep the testes from overheating, but it creates a new set of problems. Not only are they more vulnerable to damage or attack by rivals due to their exposure, but in cold weather they have to be protected against getting too cold. The biggest problem, however, is that their descent through the abdominal wall creates a weakness in the peritoneal membrane, which can lead to an inguinal hernia, a very painful and debilitating condition. Most primates walk on all fours, and the weight of their abdomen hangs down from their spine and is spread across their belly, so this is not much of a problem. But when humans began to walk upright, the weight of our abdominal organs pushed straight down and was borne in our hip region, creating a lot of pressure on the peritoneum—the ideal condition for hernias.

The weight of our entire upper body on our hips and lower spine also causes many lower back problems. Our lower spines were originally adapted for a quadrupedal gait, and the spine had to bear only part of our weight. Because our spine is not well engineered to hold the entire weight of our body in a bipedal posture, many people develop back pain and spinal problems such as herniated discs and spinal curvature (scoliosis).

Another example of our crummy design is familiar to all of us who suffer from colds, allergies, and sinus headaches. Our nasal sinuses are large cavities in our facial region that take the air we inhale through the nose and pass it over damp mucus membranes, which removes the dust and dirt, modifies its temperature, and humidifies it before it reaches our lungs. In an animal with a long snout and a quadrupedal posture, such as a dog or a baboon, the sinuses are long and have ducts that drain them at the bottom of each sinus. But one of the consequences of our upright posture is that the drainage ducts are rotated from the bottom of the chamber to the side or even to the top of the chamber, especially in the maxillary sinus, which is directly beneath your cheekbones (figure 21.1). Not only that, but in shortening our snouts, the sinuses have become flattened and narrow and don't

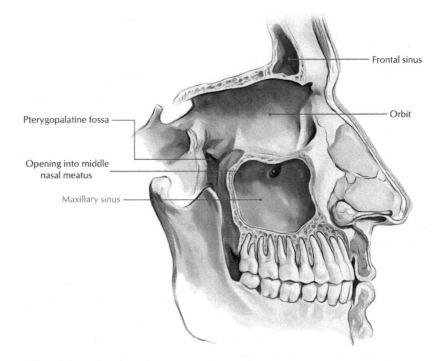

drain as well. Consequently, the mucus draining from the sinuses in our face needs to travel uphill against gravity to reach the exit duct, which leads to lots of sinus drainage problems and frequent sinus infections.

Another example of poor design can be found in the ridiculous looping path the urethra takes in the male reproductive system. It passes through the middle of the prostate gland, which can become inflamed or cancerous. When the prostate gland swells, it pinches off the urethra, making it more difficult for urine to flow from the bladder to the penis. As older men who have experienced prostate problems know, this can be very painful and force men to urinate over and over again, especially at night. Prostate cancer is the number one killer of men over 50, making this region particularly vulnerable. This looping path is a consequence of our embryonic development and of the descent of the testes, both constraints that make it poorly designed for what it does.

The rest of the reproductive system is not much better. The fallopian tubes in human females are unnecessarily long, sometimes leading to tubal or ectopic pregnancies. The opening of the fallopian tubes does not even directly connect to the ovaries, so on rare occasions a fertilized egg can escape into the body cavity and cause problems (known as an abdominal pregnancy). The vas deferens in males is also much too long, taking a long looping path rather than a shorter more efficient path. Nathan Lents spends a whole chapter in *Human Errors* listing a series of other problems in our reproductive systems that are found in no other animal and that make humans much less fertile than they otherwise could be. No other animal has such a high rate of failed pregnancies, birth defects, missed reproductive opportunities, and other problems.

The biggest problem in reproduction is that the huge expansion of the brain in human babies makes their head too large to pass through the bony opening in the hips of females that is part of the birth canal. If human females had larger pelvic openings to accommodate this brain expansion, their hips would be so wide that walking in a bipedal posture would be much more difficult. Instead, humans circumvent the problem by birthing babies prematurely (nine months of gestation versus the longer time expected for a mammal of our body size). This allows the premature baby to pass through the birth canal with its smaller brain (although it is a tight squeeze and many women have died in childbirth because it is such a problem). The baby then completes its brain growth outside the womb. The tradeoff is that the baby is so underdeveloped that he or she requires a lot more parental care after birth. Babies of mammals with normal gestation periods are much more able to take care of themselves soon after birth. However, some in the anthropological community question whether this is true or not.

The clumsy design of our bipedal limbs is due to ancestors that had a quadrupedal gait, so the hips, knees, ankles, and feet did not need to bear that much weight. In bipedalism, all of these joints bear the entire weight of our body, which is magnified when we exert extreme force on those joints by jumping or running too much. As people get older, our hips, knees, ankles, and feet give us the most problems in life because they were never designed to take so much pounding. Athletes (especially those in rough sports such as football and basketball, where these big athletes are moving very fast and making rapid changes in direction) suffer the most injuries in these joints, especially in our poorly designed knee. Many an athlete has ended a career early because of a blown knee joint (especially tearing the ligaments in the knees), a ruptured Achilles tendon in the ankle, or bad foot problems. In fact, humans are very poorly designed for bipedal running and walking, as our foot problems in particular demonstrate. If you want to see a well-designed bipedal vertebrate, look at the ostrich, which can run much faster and never has the foot or joint problems humans endure.

Or consider the ridiculous paths that some nerves take. Chapter 15 illustrated the path of the left recurrent laryngeal nerve, which runs from the upper spine to the voice box. It takes an extremely long detour in giraffes and is almost as poorly designed in humans (see figure 15.4). This nerve is associated with the sixth gill arch in our embryonic development, which was not a problem in a fish that has no neck. But as our neck developed, the left recurrent laryngeal nerve (associated with the gill arch that becomes the aorta region) must go down the neck, loop around the aorta, then come back up the neck to reach our voice box (figure 21.2). This isn't quite as ridiculous as the 15-foot detour of that nerve in the giraffe neck, but it is still a poor design. Another example is the route of the sciatic nerve in our hip and spine. As it is currently configured, it is highly prone to being trapped between bony parts of the lower spine and hip, causing the painful nerve condition called sciatica.

Anyone who has accidentally inhaled food has experienced the poor design of our throat cavity. We have just one little valve—the epiglottis—to keep food or fluids from going into our trachea and lungs rather than into our esophagus, so we cannot swallow and take a deep breath at the same time (figure 21.3). If we swallow when the epiglottis is not completely closed, we inhale food or fluids and choke—sometimes to death if we don't cough it out or if no one is handy to perform the Heimlich maneuver. Almost 5,000 Americans choked to death in 2014 alone, mostly due to inhaling food or drink. It doesn't help that our instinctual reaction to surprise is to gasp and inhale suddenly, which sucks food or fluid into our lungs. Surely this is a system that could have been better engineered. Most animals do not share this problem; they have completely separate systems for breathing and eating that do not interfere with each other. Our system was inherited from an embryonic system that developed lungs and the digestive tract in a certain configuration, and the clumsy setup is not detrimental enough for

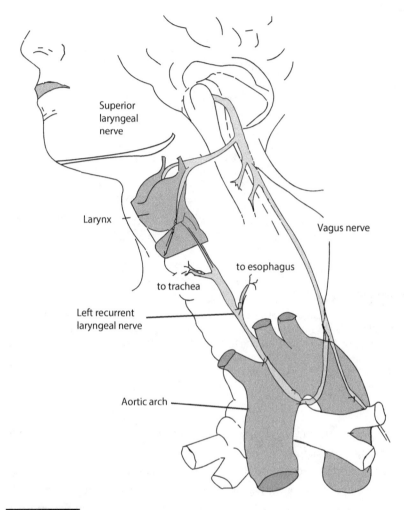

Figure 21.2 ▲

The indirectly and poorly designed wiring of the left recurrent laryngeal nerve in humans, which begins at the voice box, loops down into the chest cavity around the aorta, then comes back up the neck to the spinal cord, in a circuitous path, when it could be directly linked with just a few inches of nerve cord. (Redrawn from several sources by E. Prothero)

selection to override embryology and design a better system. This system is much worse for humans than it is for other animals because, in developing speech, we moved the voice box much higher in the throat than it is in other animals—making our throat and trachea configuration more prone to choking us to death. This is particularly true for babies, whose muscular coordination of the throat and epiglottis are not yet fully developed.

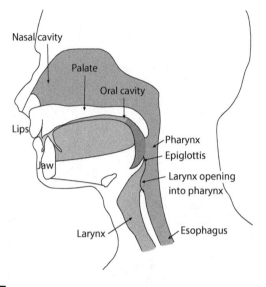

Figure 21.3 ▲

The complicated path of the digestive tract and respiratory system in the human head and neck. When we are swallowing, the flap of flesh called the epiglottis is supposed to close over the opening to the trachea and larynx, preventing food or liquid from going into our lungs. However, all too often something goes wrong, and we inhale food or liquid and begin to choke. (Redrawn from several sources by E. Prothero)

Their very short throat passages and shallow throats make them even more vulnerable to choking.

Most of these poorly designed systems are a consequence of embryology and developmental pathways, or of superimposing a bipedal gait on a body that inherited a quadrupedal gait. But just as striking are examples of organ systems that no longer have any function and are truly rudiments or vestigial systems. They range from the simple example of goose bumps to more complex systems. Goose bumps are caused by the muscles around the roots of our body hairs contracting to raise the hairs up and fluff them out, either to create an insulating layer when it's cold or to register fear (as many animals do when they fluff out their fur or feathers to appear larger during a time of a threat). Most humans have little body hair, and the muscles around the hair follicle just form little goose bumps instead, which are completely useless.

In 1893, anatomist Robert Wiedersheim published *The Structure of Man: An Index to His Past History*, which listed at least 86 human organs that are considered vestigial (later expanded to 180 examples). Biologist Horatio Newman, testifying in favor of evolution during the Scopes trial in 1925, wrote that "there are, according to Wiedersheim, no less than 180 vestigial structures in the human body, sufficient to make of a man a veritable walking museum of antiquities." Some of these have since been shown to have at least a minimal function, but most are truly useless rudiments of once-functional systems. As Wiedersheim points out, a vestigial organ need not be totally useless, but only "wholly or in part functionless." The use for these organs has greatly diminished if not vanished or, in Wiedersheim's words, "lost their original physiological significance." Evolution deniers try to salvage this hopeless situation by pointing to some vestigial organs that have a tiny bit of function left in an attempt to discredit this entire line of evidence. But this misses the point. If the organ system is greatly reduced (compared to the ancestral condition) and performs minimal function, it is still evidence of a past functional system that has degenerated and thus is not well designed.

One of the most famous of these systems is the appendix, a redundant organ that in herbivorous mammals provides enzymes to the intestine to aid in digesting cellulose. But it no longer functions this way in humans because we are not primarily herbivores and don't have the specialized gut bacteria to efficiently digest cellulose. For a long time the appendix was touted as a classic vestigial organ, and it is certainly a problem for humans when it becomes infected. If the appendix bursts, it can kill the bearer of the infection. More recent research suggests that it may supply some of the bacteria to the gut flora, especially if the gut flora has been wiped out by antibiotics. But that small benefit has to be weighed against the huge life-threatening risk of appendicitis or a burst appendix.

Another example is our ridiculously small tailbone. All monkeys and more primitive primates have a long tail for balance and other functions (some even have prehensile tails for grasping limbs), as do most mammals. We humans also had a long tail when we were embryos (see figure 5.3). But this tail-making gene is shut off during embryology in all apes and humans, and our early embryonic tail is resorbed. Instead of being born with a fully functional tail, all we have is three tiny tailbones at the end of our spine (the coccyx). A few tiny muscles still insert in the tail region, so it is not entirely functionless, but the fact that it's reduced to a tiny stub shows that its function is relatively unimportant now. The genes for the ape and human tail are not lost. Every once in a while gene regulation fails, and the gene to shut off

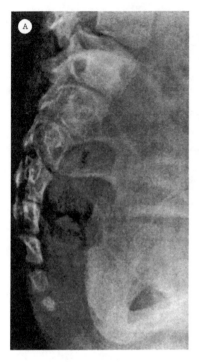

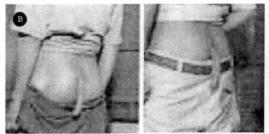

Figure 21.4 ▲

Every once in a while a human is born with an atavistic tail, a throwback to our evolutionary past when the regulation that normally shuts down our genes for tails fails to operate. The human tail comes complete with fully developed vertebrae, muscles, and other features of animal tails. (*A*) X-ray of a human with well developed tail vertebrae. (*B*) Image of two humans with fully developed tails. (From J. A. Bar-Maor, K. M. Kesner, and J. K. Kafton, "Human Tails," *Journal of Bone and Joint Surgery* 62-B, no. 4 [November 1980], 508–510; used with permission)

tail development doesn't work. In that instance, the human develops a full-fledged tail, complete with much larger vertebrae in the coccyx (figure 21.4). This is a famous example of an atavism, or evolutionary throwback, where long-lost features suddenly reemerge. Any time you want to deny that you are a monkey's cousin, just remember you have the genes for a monkey's tail.

Look at the problem some people have with their "wisdom teeth," the third and last molars in their jaws (figure 21.5). Most of our primate relatives have relatively long snouts and jaws, and they have room for all of their teeth. Early humans had prominent snouts and used their large molars for eating gritty vegetation. But modern humans have very short faces and snouts with a shortened tooth row, so we hardly have room any more for all of our teeth. Some time in our late teens, as our last molars finally begin to erupt, often there is not enough room in our short jaws for them to fully emerge. They become impacted or otherwise cause problems, and often require surgical removal. Wisdom teeth are not only useless but could result in disease or death if they become damaged or deformed or infected. The wisdom teeth are so useless now that many human populations (such as the

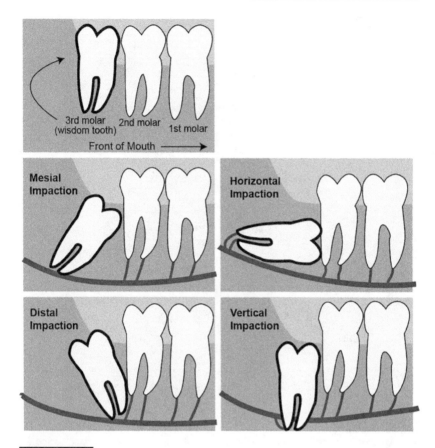

Figure 21.5 ▲

Instead of the ancestral snout that apes have, the human face and tooth row is now so short that there is often not enough room to allow the last molars ("wisdom teeth") to erupt fully. This often results in impacted or crooked wisdom teeth, and many people have to have them surgically removed. In some individuals, the teeth never even develop or erupt at all, so they are gradually being lost in human populations. (Redrawn from several sources by E. Prothero)

Aborigines from Tasmania) never erupt them at all, thanks to the activity of the *pax-9* gene, which controls eruption sequence. Given enough time, wisdom teeth will completely vanish from all human populations.

Even more striking is the external ear. Our monkey relatives (such as macaques) have an ear that can swivel and rotate in any direction to pick up sound around them. Watch a horse or a cow or a cat or a short-eared dog move their ears to track sounds. They can do this even if they are napping or looking in a different direction. Our ape relatives have the same muscles

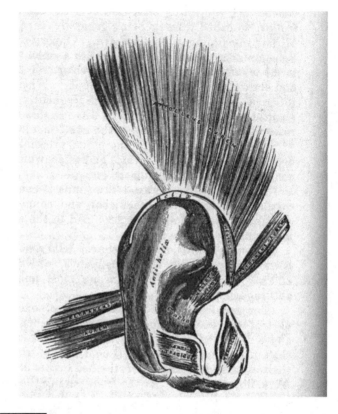

Figure 21.6 ▲

Although most humans cannot wiggle their ears at all, or move any of their ear muscles, we still retained these vestigial organs even though they no longer perform any useful function. (Courtesy of Wikimedia Commons)

for the ear but have limited ability to move their external ears. Our external ears are almost entirely immobile and pressed against the side of our head. In most of us, the ear muscles are still there, although they are tiny and functionless (figure 21.6). Just a tiny percentage of humans have ear muscles strong enough to wiggle their ears slightly, but the function of ear muscles to turn our ear toward a sound source is completely gone. One reason we no longer have mobile ears is that apes and humans can easily turn their entire head in the horizontal plane, something that most mammals cannot do as easily. Thus we have developed a new functional ability to move our head to track sound and lost the ability to use our ears for that function.

This list could go on and on (see the books by Held or by Lents on the Further Reading list for many more examples). If we were specially created and ideally designed, why do we have tiny remnants of the nictitating membrane in the corner of the eye? This membrane serves to cover and protect the eyeball in many other mammals and other vertebrates, but in primates only a tiny vestige remains as a functionless fold of tissue in the inside corner of the eye. Many animals have a vomeronasal organ in their palate, which serves as a secondary source of smell. But in humans the vomeronasal organ is gone or is just a tiny vestige, and the nerve bundles that serve it have degenerated and no longer function; even the genes that make this organ have become nonfunctional pseudogenes. Babies still have a powerful palmar grasp reflex, which is used to hang onto the fur of their mothers—but human mothers are usually not hairy, so this reflex no longer serves a useful function. Or consider the long list of nonfunctional genes and junk DNA that all humans carry (see chapter 18).

Here is a simple final example: If humans were specially created and well designed, why do men have nonfunctional nipples? It makes sense in evolutionary terms because the nipples and other secondary sexual features are not strongly determined in embryos, and only after they are stimulated by estrogen or testosterone does the fetus begin to develop all the distinctively different male and female sexual organs. But in embryos, the genes for nipples are always present, even if males do not develop large mammary glands to nurse their babies as females do. And in some humans the developmental pathway even goes further astray and makes extra nonfunctional nipples in males—as many as seven! This simply adds to the gigantic pile of evidence that humans are not well designed; we are a pile of semifunctional and nonfunctional organs that stand as silent witnesses to our evolutionary past.

FOR FURTHER READING

Dao, Anh H., and Martin G. Netsky. "Human Tails and Pseudotails." *Human Pathology* 15, no. 5 (1984): 449–453.

Fallon, John F., and B. Kay Simandl. "Evidence of a Role for Cell Death in the Disappearance of the Embryonic Human Tail." *American Journal of Anatomy* 152, no. 1 (1978): 111–129.

Held, Lewis I., Jr. *Quirks of Human Anatomy: An Evo-Devo Look at the Human Body.* Cambridge: Cambridge University Press, 2009.

Lents, Nathan H. *Human Errors: A Panorama of Our Glitches, from Pointless Bones to Broken Genes.* New York: Houghton Mifflin Harcourt, 2018.

Mukhopadhyay, Biswanath, Ram M. Shukla, Madhumita Mukhopadhyay, Kartik C. Mandal, Pankaj Haldar, and Abhijit Benare. "Spectrum of Human Tails: A Report of Six Cases." *Journal of Indian Association of Pediatric Surgeons* 17, no. 1 (2012): 23–25.

Pereira, Tiago V., Francisco M. Salzano, Adrianna Mostowska, Wieslaw H. Trzeciak, Andrés Ruiz-Linares, José A. B. Chies, Carmen Saavedra, Cleusa Nagamachi, Ana M. Hurtado, Kim Hill, Dinorah Castro-De-Guerra, Wilson A. Silva-Júnior, and Maria-Cátira Bortolini. "Natural Selection and Molecular Evolution in Primate PAX9 Gene, a Major Determinant of Tooth Development." *Proceedings of the National Academy of Sciences* 103, no. 15 (2006): 5676–5681.

Saraga-Babić, Mirna, Eero Lehtonen, Anton Švajger, and Jorma Wartiovaara. "Morphological and Immunohistochemical Characteristics of Axial Structures in the Transitory Human Tail." *Annals of Anatomy—Anatomischer Anzeiger* 176, no. 3 (1994): 277–286.

Shubin, Neil. *Your Inner Fish: A Journey Into the 3.5 Billion-Year History of the Human Body.* New York: Vintage Books, 2009.

Spiegelmann, Roberto, Edgardo Schinder, Mordejai Mintz, and Alexander Blakstein. "The Human Tail: A Benign Stigma." *Journal of Neurosurgery* 63, no. 3 (1985): 461–462.

Spinney, Laura. "Remnants of Evolution." *New Scientist* 198, no. 2656 (2008): 42–45.

Wiedersheim, Robert. *The Structure of Man: An Index to His Past History.* London: Macmillan, 1893.

THE THIRD CHIMPANZEE

We must, however, acknowledge, as it seems to me, that man with all his noble qualities, still bears in his bodily frame the indelible stamp of his lowly origin.

—CHARLES DARWIN, *THE DESCENT OF MAN* (1871)

The next time you visit a zoo, make a point of walking by the ape cages. Imagine that the apes had lost most of their hair, and imagine a cage nearby holding some unfortunate people who had no clothes and couldn't speak but were otherwise normal. Now try guessing how similar those apes are to us in their genes. For instance, would you guess that a chimpanzee shares 10 percent, 50 percent, or 99 percent of its genetic program with humans?

—JARED DIAMOND, *THE THIRD CHIMPANZEE* (1992)

One fundamental discovery about humans is how closely related we are to the other great apes. As early as 1735, the founder of modern classification, Carolus Linnaeus, placed humans within the apes, monkeys, and lemurs as Primates, and diagnosed our species with the old Greek phrase *Gnothe sauton* (Know thyself). In 1766, the Count of Buffon wrote in volume 14 of his *Histoire Naturelle* that an ape "is only an animal, but a very singular animal, which a man cannot view without returning to himself." Other French naturalists, including Cuvier and Geoffroy, commented on the extreme anatomical similarity of apes and humans, although they refused to say that humans were a kind of ape. The pioneering French biologist Lamarck explicitly argued in *Philosophie Zoologique* in 1809:

> Certainly, if some race of apes, especially the most perfect among them, lost,
> by necessity of circumstances, or some other cause, the habit of climbing

trees and grasping branches with the feet, . . . and if the individuals of that race, over generations, were forced to use their feet only for walking and ceased to use their hands as feet, doubtless . . . these apes would be transformed into two-handed beings and . . . their feet would no longer serve any purpose other than to walk.

Charles Darwin certainly appreciated the close similarity between apes and humans and was interacting with the baby orangutan Jenny only a year after he returned from his *Beagle* voyage (see chapter 23). When it came time to publish, however, Darwin knew that most people were not ready to accept the idea that we were closely related to apes, or even worse, descended from an ape-like ancestor. He tiptoed around the problem in his revolutionary 1859 book, *On the Origin of Species*, saying only that "light will be thrown on the origin of man." Of course, the people at the time were not fooled and decried his ideas that "men came from monkeys" and accused him of blasphemy and heresy. It wasn't until 1871 that Darwin dealt with the issue directly in *The Descent of Man*. His main supporter and advocate Thomas Henry Huxley was not so timid. As early as 1863, in *Zoological Evidences of Man's Place in Nature*, Huxley boldly showed the extreme similarities of the skeletons of humans and great apes, which differ only in the proportions of their limbs and the shape of the skull (see figure 12.2B). Humans are bone for bone identical with the rest of the apes, differing only in size and proportions. The arm-swinging apes (gibbons and orangutans) have much longer arms than do humans, and humans have the longest legs of all the apes.

The gulf between humans and the rest of the apes has narrowed considerably over the years. Instead of the old "screaming hooting ape" stereotype, we have discovered just how similar apes are to humans. Decades of field research by pioneering anthropologists Jane Goodall with the chimpanzees, Birute Galdikas with orangutans, and the late Dian Fossey with the mountain gorillas have demystified these majestic creatures and surprised us with their amazing behavioral similarities to humans (see chapter 23). Both chimpanzees and gorillas can learn sign language, communicate in simple sentences, and make and use simple tools. Their societies are very sophisticated compared with those of any other animal and provide many insights into the complexities of human societies as well. Over a century of research by hundreds of anthropologists has documented more and more connections between apes and humans. The boundary between "human" and "animal" has all but vanished: Every time someone has tried to prove we are more "special" than any animal, the anatomical or behavioral evidence shows that at least some animals have that feature as well. As a result, the entire scientific community and most educated people now accept that humans are part of nature and are connected to other animals in many profound ways.

If the fossil record of human evolution is not evidence enough (see chapter 24), the clinching argument is found in every cell in your body. Unbeknownst to Darwin or any other biologist before the 1960s, another source of data clearly shows our relationship to apes and the rest of the animal kingdom: the structure of our DNA and other biomolecules, such as proteins. Even the very first molecular techniques demonstrated that our DNA and chimp and gorilla DNA were extremely similar. When you put the serum of antibodies of humans and apes in the same solution, the reactions are much stronger than between humans and any other animal, suggesting that the immunity genes of humans and apes are most similar. This is called the immunological distance method for estimating our relatedness to other organisms. Then, in the late 1960s, a technique called DNA-DNA hybridization showed that chimp DNA is virtually identical to ours.

In the past 30 years, technological leaps including the polymerase chain reaction (PCR) have made it possible to directly sequence the DNA not only of humans but also of many other animals and plants. Geneticists sequenced the entire DNA of humans first in 2000 and the chimp DNA in 2005. When they are compared, we get the same result—humans and chimps share 98-99 percent of their DNA. Less than 2 percent of our DNA differentiates us from chimps and from gorillas as well. Remember that 60 to 90 percent of our DNA is noncoding (sometimes referred to as junk DNA) and is never read or used but is passively carried from generation to generation (see chapter 18). Some of this junk is endogenous retroviruses (ERVs), which are remnants of viral DNA that was inserted into our genes when some distant ancestor was infected, and it is still carried around even though it no longer codes for anything.

One of the pseudogenes humans have is an inactive copy of the GULO gene, which manufactures vitamin C. Apparently, our primate ancestors ate so much fruit that they no longer needed to make their own vitamin C, and this gene was shut off without harming them. (The only other group that survives without making its own vitamin C are fruit bats.) When humans don't get enough vitamin C, they can get a nasty disease called scurvy. Sailors used to suffer from bleeding gums, lost teeth, and many other symptoms of scurvy on a regular basis because of their limited diet at sea, which did not include fresh fruit or vegetables. The British Royal Navy eventually discovered that citrus fruits, such as limes, provided this essential vitamin and routinely added them to sailors' diets (which is why British sailors are called "limeys"). We must eat more fruit or take vitamin pills to replace this essential vitamin because our vitamin C gene can no longer be switched back on so we could make it ourselves.

In fact, our bodies can no longer manufacture a whole sequence of "essential nutrients," and we need to consume them in our diet to prevent suffering from various vitamin deficiencies. Deficiencies in B vitamins, for example, cause a range of diseases in humans but almost no wild animals suffer from them. Lack of B1 (thiamine) causes beriberi; lack of niacin (B3) causes pellagra; lack of B6 (pyridoxine) causes neurological disturbances; absence of B9 (folate) causes birth defects and anemia; and a shortage of B12 (cobalamin) produces macrocytic anemia. Many animals either have the genes to make these nutrients themselves or get them from their diets, but we cannot produce them, and if our diet is low in any of these, we suffer the results.

Besides the true junk DNA, a smaller percentage of the genome consists of structural genes that code for every protein and structure in our body, including genes we no longer use. The 1 to 2 percent that distinguishes us from chimps are mostly regulatory genes, the on/off switches that tell the rest of the genome when to express a particular gene. These genes are the reason humans look so different from other apes, even though our genes are nearly identical.

When we look in detail at that remaining 1 to 2 percent that differentiates us from chimps, what do we find? As neurologist and biological anthropologist Robert Sapolski pointed out, about half of that different DNA is for the olfactory region: chimps have a much more sensitive sense of smell than we do. Many of the genes for our sense of smell are turned off and have become pseudogenes. Another difference is in the genes for the development of the hip and thighbones, which allows us to walk upright easily, whereas chimps have an awkward walking gait. These are the genes that make the biggest difference in our skeleton and posture. A few genes code for body hair, which chimps have all over and most humans don't. Another codes for differential responses to diseases, so we can survive tuberculosis that is rapidly fatal in chimps, but chimps are resistant to simian AIDS whereas its human counterpart (which originally jumped from apes to humans) is fatal in humans. The biggest genetic difference between humans and chimps and apes is in the genes that code for creating neurons. Our brain structure is fundamentally the same as that of chimps and most mammals, but chimp brain neuron development stops at a particular level, whereas human genes allow for additional rounds of neuron development, producing our unusually large brains. From this tiny part of our genome, not only do we produce thousands of extra neurons compared to other animals but also everything from music to art to philosophy to science to religion, plus everything the human brain creates that no other animal brain can envision.

The extreme similarity of our genes to those of the two species of chimpanzee (the common chimp, *Pan troglodytes*, and the pygmy chimp or bonobo, *Pan paniscus*) should, all by itself, be overwhelming and convincing evidence of our close relationship. Despite some people's aversion to the idea, we are indeed the ape's reflection. Biologist Jared Diamond puts it this way: Imagine that some alien biologists came to Earth and the only samples they could obtain were DNA. They sequenced many different animals, including humans and the other two chimps. Based on these data alone, they would conclude that humans are just a third species of chimpanzee. Our DNA is more similar to that of the other two chimps than the DNA of most species of frog is similar to one another. We and the other chimps are even more similar than lions and tigers are to each other, which share about 95 percent of their DNA and can interbreed in zoos. The differences in appearance between apes and humans are caused by tiny changes in the regulatory genes, which produce huge results. The evidence from our genes, as well as from our anatomy, is overwhelming. The DNA in every cell in your body is a testament and witness to your close relationship to chimps.

In summary, the 1 to 2 percent of the genome that differentiates us from chimps are the result of regulatory genes that turn structural genes on and off (which make up most of the 98 percent of the genome that is the same). We still have the genes for most parts of the ape body, and the monkey body too, and every once in a while a genetic mistake, or atavism, occurs and humans express long-repressed genes that we still carry (for example, to make a tail; see chapter 21).

Since the 1920s, many biologists and anthropologists have argued that much of what differentiates us from the chimpanzee is a well-known phenomenon in nature called neoteny (literally, "holding on to youth"). Many animals in nature find ways to reproduce and have offspring while their bodies are still in their juvenile form. For example, certain salamanders known as axolotls (genus *Ambystoma*) that live in lakes in Mexico are able to reproduce while they still have their juvenile gills. But if the lake water goes bad or dries up, they complete their development to adult salamanders, lose their gills and develop lungs, and then crawl off to some other pond. The ability to change the timing of reproduction with respect to development gives animals great flexibility to take advantage of their existing genetic instructions without having to make a drastic genetic or evolutionary change.

If you look at a juvenile chimpanzee (figure 22.1A), its skull is much like that of a human, with a large brain, small brow ridges, short snout, and upright posture. During development into an adult (figure 22.1B), the

Figure 22.1 ▲

Neoteny in apes and humans. (*A*) Juvenile chimpanzees have many characteristics found in adult humans, including an upright posture, a relatively large brain, small brow ridges, and a less protruding snout. (*B*) As they grow into adult chimps, these features all become more ape-like. Since the 1920s, many anthropologists have argued that much of what makes us human is retention of juvenile ape characteristics into adulthood. (From Adolf Naef, "Über die Urformen der Anthropomorphen und die Stammesgeschichte des Menschenschädels," *Naturwissenschaften* 14 [1926]: 445–452)

chimpanzee develops the larger snout with long canines, big brow ridges, and forward slouching posture of the head. If regulatory genes tweak our embryonic development a tiny bit, most of the characteristics that mark us as human can be expressed just by remaining juvenile apes, reaching sexual maturity without every truly growing up.

You may have read some of the fascinating books by Aldous Huxley, novelist and author of the dystopian classic *Brave New World* (a high school reading list favorite). His brother was the famous evolutionary biologist Julian Huxley, and both were grandsons of Darwin's biggest supporter, Thomas Henry Huxley. Aldous knew these ideas about human neoteny very well because of his brother's influence. In 1939 in his novel *After Many a Summer Dies the Swan*, Aldous Huxley explored the idea of immortality and the human desire to find a way to extend life beyond what nature intended. The main character is millionaire Jo Stoyte (modeled after legendary press baron William Randolph Hearst, whom Huxley met when he was a Hollywood screenwriter in the 1920s), who is attempting to buy eternal life by hiring a classic "mad scientist" character, Dr. Obispo, to do research on delaying aging and prolonging life. Obispo discovers that the Third Earl of Gonister in England had lived several centuries without any signs of aging, apparently by ingesting carp guts. Archival records showed that he had fathered children when he was over 100 years old. In a plot twist, Obispo rapes the millionaire's mistress (modeled on Hearst's real mistress, actress Marion Davies), and the millionaire accidentally kills Obispo's scientific assistant in a jealous rage. Stoyte and Obispo have to run from the law, so they flee to England and try to discover what happened to the Third Earl of Gonister. Finally, they break into his castle and find him in the basement, still alive and 201 years old—all grown up into an adult ape.

FOR FURTHER READING

Diamond, Jared. *The Third Chimpanzee: The Evolution and Future of the Human Animal.* New York: HarperCollins, 1992.

Harris, Eugene E. *Ancestors in Our Genome: The New Science of Human Evolution.* Oxford: Oxford University Press, 2015.

Jurmain, Robert, Lynn Kilgore, Wenda Trevathan, Russell L. Ciochon, and Eric Bartelink. *Introduction to Physical Anthropology.* 15th ed. New York: Cengage Learning, 2017.

Lewin, Roger, and Robert A. Foley. *Principles of Human Evolution.* 2nd ed. Malden, Mass.: Wiley-Blackwell, 2003.

Marks, Jonathan. *Human Biodiversity: Genes, Race, and History.* London: Routledge, 2017.

——. *What It Means to Be 98 percent Chimpanzee: Apes, People, and Their Genes.* Berkeley: University of California Press, 2002.

Potts, Richard, and Christopher Sloan. *What Does It Mean to Be Human?* Washington, D.C.: National Geographic, 2010.

Reich, David. *Who We Are and How We Got Here: Ancient DNA and the New Science of the Human Past.* New York: Pantheon, 2018.

Roberts, Alice. *Evolution: The Human Story.* London: DK Books, 2018.

Rutherford, Adam. *A Brief History of Everyone Who Ever Lived: The Human Story Retold Through Our Genes.* New York: The Experiment LLC, 2016.

Shipman, Pat. *The Animal Connection: A New Perspective on What Makes Us Human.* New York: Norton, 2011.

Sibley, Charles G., and Jon E. Ahlquist. "The Phylogeny of Hominoid Primates, as Indicated by DNA-DNA Hybridization." *Journal of Molecular Evolution* 20 (1984): 2–15.

THE APE'S REFLECTION

No one who looks into a gorilla's eyes—intelligent, gentle, vulnerable—can remain unchanged, for the gap between ape and human vanishes; we know that the gorilla still lives within us. Do gorillas also recognize this ancient connection?

—GEORGE SCHALLER

People had noted the similarities between humans and apes going back to before the days of Linnaeus (see chapter 22). When Darwin's *On the Origin of Species* came out in 1859, people were especially shocked and appalled by the idea that they were closely related to the hairy hooting apes they saw in the zoo or circus (figure 23.1). Everyone was in deep denial, trying to find things that clearly separated us from the rest of the animal kingdom.

In 1857 and 1858, before Darwin's book was published, the eminent naturalist Richard Owen, Darwin's most determined critic, anticipated the issue. He tried to argue that humans were special and distinct from apes because we possessed a structure in the brain called the hippocampus that apes did not have. The name "hippocampus" means "seahorse," and it is shaped a bit like a seahorse. (We now know this structure helps the two halves of the brain communicate with each other and converts short-term memory to long-term memory.) In 1860 and 1861, as the debate over evolution was heating up, Darwin's staunchest supporter, Thomas Henry Huxley, challenged Owen's declarations. Huxley dissected a gorilla and chimps and other primates and found that they all had a hippocampus. The debate continued to be the hottest scientific controversy in Britain in 1862, but many

Figure 23.1 ▲

Many contemporary editorial cartoons mocked the idea that humans were related to apes, (A) showing Darwin as an ape or (B) showing Darwin and an ape looking in the mirror to see their resemblance to each other. (Courtesy of Wikimedia Commons)

other scientists soon confirmed Huxley's work with their own dissections. Within a few years, the hippocampus ceased to be a point of contention because the anatomy was fundamentally the same. Naturalists retreated from this argument and began to define the specialness of humans by their tool use, symbolic language, and other supposedly unique characteristics.

The insulting idea that we might be related to apes came up in many debates, especially the famous Huxley-Wilberforce debate in 1860. Huxley had been championing Darwin's controversial new book for months. A pro-Darwinian talk on June 27, 1860, at the meeting of the British Association for the Advancement of Science held in Oxford, was challenged by Richard Owen. Then rumors spread that the Bishop of Oxford, "Soapy Sam" Wilberforce (so named because he was slippery and hard to pin down in debates), would attend the meeting and give a speech the next Saturday. Huxley had planned to leave the meeting, but Robert Chambers persuaded him to stay. Wilberforce had no real understanding of what he was talking about, but he had been coached by Owen (his childhood friend) and supplied with lots of supposed problems with evolution. Wilberforce's speech on June 30, 1860, was good-humored and witty, but it was an unfair attack on Darwinism, ending in the now infamous question to Huxley of whether "it was through his grandfather or grandmother that he claimed descent from a monkey." At this insult, Huxley is purported to have said to chemistry Professor Brodie seated next to him, "The Lord has delivered him into mine hands."

When Huxley got up to speak, he responded that he had heard nothing from Wilberforce to prejudice Darwin's arguments, which still provided the best explanation of the origin of species yet advanced. He ended with the equally famous response to Wilberforce's question, that he had "no need to be ashamed of having an ape for his grandfather, but that he would be ashamed of having for an ancestor a man of restless and versatile interest who distracts the attention of his hearers from the real point at issue by eloquent digression and skilled appeals to religious prejudice." Allegedly, the crowded hall broke into a riot, women fainted, and nothing more could be heard because of the screaming and shouting from both sides. Unfortunately, no one recorded the exact words of the debate, so much of this is based on the imperfect accounts of those who were present.

Charles Darwin certainly appreciated the similarity of apes and humans, and it had a strong influence on his ideas. As early as 1838, one year after returning from his landmark voyage around the world on the HMS *Beagle*, the young naturalist visited the London Zoo and saw their baby orangutan named Jenny. He spent many hours with her in her cage, noting her emotions, her behaviors, and how similar these were to humans in so many ways. In addition, he studied two of the zoo's other orangutans. In his unpublished notebooks, Darwin wrote extensively about their apparent tool use and creative play in fashioning their own toys out of sticks.

Writing about Jenny, Darwin noted:

She is fond of breaking sticks & in overturning things to do this (& she is quite strong) she places tries the lever placing stick in hole & going to end as I saw.— She will take the whip & strike the giraffes, & take a stick & beat the men.— When a dog comes in she will take hold of anything, the keepers say, decidedly from knowing she will be able to hurt more with these than with paw.

Most impressive were the orangutans' challenges with self-awareness when given a mirror. As Darwin's notes describe it:

> Both were astonished beyond measure at looking glass, looked at it every way, sideways, & with most steady surprise.—after some time stuck out lips, like kissing, to glass, & then the two did when they were first put together.—at last put hand behind glass at various distances, looked over it, rubbed front of glass, made faces at it—examined whole glass—put face quite close & pressed it—at last half refused to look at it—startled & seemed almost frightened, & evidently became cross because it could not understand puzzle.—Put body in all kinds of positions when approaching glass to examine it.

In 1839, when Darwin's first child, William, was born, Charles did the same types of experiments, analyzing the child's emotions and expressions and developmental landmarks and comparing them to apes. He continued the tradition of scientifically observing his children with his first daughter, Anne Elizabeth, when she was born in 1841. He lost interest in this project for the eight later children that his wife Emma bore him (three of whom died in childhood). Darwin was deeply impressed with how many human emotions and behaviors the great apes showed, and he got past the superficial sneering stereotype of hooting subhuman behavior that most people believed at the time.

Nonetheless, most people recoiled at the idea of humans being close kin of apes and monkeys, and phrases like "Well, I'll be a monkey's uncle" refers to this revulsion. When protesters attacked evolutionary thinking during the Scopes "Monkey Trials" in the summer of 1925, not only did the nickname for the trials reflect this obsession with our relation to primates but statements of the principals and even signs carried by the crowds also reflected this horror at being considered closely related to apes and monkeys. During the carnival atmosphere of the trial in Dayton, Tennessee, an exhibit featuring two chimpanzees and a supposed "missing link" opened in town, and vendors sold bibles, toy monkeys, hot dogs, and lemonade. The missing link was in fact Jo Viens of Burlington, Vermont, a 51-year-old man who was of short stature and possessed a receding forehead and a protruding jaw. One of the chimpanzees—named Joe Mendi—wore a plaid suit, a brown fedora, and white spats, and entertained Dayton's citizens by monkeying around on the courthouse lawn. All the editorial cartoons of the time (as in Darwin's days) featured some kind of image of monkeys or apes because these were the most visually striking images that evoked the idea of evolution.

The mistrial and failure of the Scopes verdict lasted for decades, mostly in the watering down or elimination of evolution in biology textbooks. Then, in 1957, the launch of Sputnik showed that American science had fallen far behind the Soviets. A new more rigorous science curriculum (including biology) was developed and returned to the public schools, and evolution was taught again across the United States. Still, monkeys and apes were viewed as silly and primitive and crude, and they played comic relief in movies and TV for many decades. The screeching, hooting chimp was a mainstay of movies from Ronald Reagan's *Bedtime for Bonzo* to numerous Disney family movies that used trained chimps to amuse the viewer. (Ironically, the toothy "grin" that chimps make when they pull back their lips and bare their teeth is not a smile, but a threat display, a sign of fear and hostility.) Meanwhile, gorillas were viewed as powerful vicious brutes who could easily tear humans limb from limb, thanks to *King Kong*, the "Tarzan" movies, and many other African adventure films in which intrepid explorers battled terrifying, monstrous gorillas.

Most of what people knew about monkeys and apes came from the artificial setting of zoos and seeing captive trained animals. Naturally, these ideas were highly distorted because these primates did not experience normal interactions with members of their own species very often, if at all. Some pioneering wildlife biologists had done limited field studies on gorillas and chimpanzees in the wild, but not much was documented over long periods of time. In 1959, the 26-year-old biologist George Schaller began his studies of the mountain gorillas of Virunga volcanoes on the border of Rwanda, Uganda, and the Democratic Republic of the Congo. Several years of field studies culminated in two books about gorillas, in 1963 and 1964, in which Schaller debunked the old myths and showed that gorillas were very gentle, intelligent, caring creatures who rarely engaged in the chest-pounding threat displays that Hollywood stereotypes had persisted in showing for decades. As Schaller wrote in a 1995 article in *National Geographic*: "No one who looks into a gorilla's eyes—intelligent, gentle, vulnerable—can remain unchanged, for the gap between ape and human vanishes; we know that the gorilla still lives within us. Do gorillas also recognize this ancient connection?" Schaller then went on to make a reputation as one of the pioneering field biologists and animal behavior experts, with years of field research on tigers, pandas, lions, snow leopards, jaguars, capybaras, and caimans. He even discovered three species of mammals that were new to science or thought to be extinct.

Other than Schaller, however, few people had spent much time studying the great apes in the wild to determine how their natural behavior patterns compared to what people had observed in captive animals. Into this breach stepped Louis S. B. Leakey, one of the giants of anthropology. In the late 1950s, after almost 30 years of searching. he was still struggling to find early human fossils in the ancient rocks of East Africa. In 1959, his wife Mary found fossils of the robust australopithecine they nicknamed "Dear Boy" but formally named *Zinjanthropus boisei* (it is now considered to be *Paranthropus boisei*). It became a sensation and cemented Leakey's reputation as one of the foremost paleoanthropologists, especially when it was dated over 1.8 million years old in 1960, making it (and human evolution) far older than anyone imagined at the time. The Leakeys were soon the darlings of *National Geographic*, and they had almost continuous funding to keep their work going long after Louis died.

Even before his big successes, Louis Leakey was known as a spellbinding speaker who could captivate audiences not only with his descriptions of his discoveries of our human origins but also with his philosophical musings about the meaning of humanity's prehistory. He frequently left Africa to go on lecture tours, which helped raise money in the days before *National Geographic* made him a big star and gave him generous support. At every stop, flocks of star-struck people mobbed him and ask for opportunities to become anthropologists like him. Since 1946, Leakey himself had been looking for recruits to study the great apes in the wild. This research interest was inspired by studying the ecology of the ancient ape fossil *Proconsul*, which Leakey had found on Rusinga Island in 1931. In 1956, he sent his secretary, Rosalie Osborn, to study the mountain gorillas, but she lasted only four months under the harsh conditions.

In 1957, a 23-year-old British woman named Jane Goodall came to Kenya to study the animals of Africa. At first she found work as a secretary, then contacted Leakey and became his secretary. Leakey was convinced that Goodall had the right stuff to be a field researcher in primates, and in 1958 Leakey sent Goodall back to England to be trained in primate behavior by Osman Hill and in primate anatomy by John Napier. Eventually, Leakey raised the funds to support her research, and on July 14, 1960, Goodall began her research on the chimps of the Gombe Stream National Park, beginning a study that she would continue for the next 55 years.

Leakey's next recruit was Dian Fossey, who broke away from her boring work in a hospital and spent her life savings in 1963 to visit Africa and find out about studying wildlife. She visited the Leakeys in Olduvai Gorge, where she tripped, sprained her ankle, fell into one of the excavations, and then vomited on a giraffe fossil. She returned home to take a job and pay back the money she had borrowed for her African trip. Leakey gave a lecture in 1966 in Louisville that Fossey attended, and after the lecture she spoke to him and he remembered her. Soon after that, he recruited her to come to the Virunga volcanoes in January 1967 and continue the studies of the mountain gorillas where George Schaller had left off almost a decade earlier. She developed tremendous empathy for the gorillas, made many important discoveries, and fought back against the poachers who were killing them off. In 1985 Fossey was murdered, possibly by the poachers who hated her.

The last of the three women (nicknamed the "Trimates" or "Leakey's Angels") Leakey recruited was Biruté Galdikas. She heard Leakey lecture at UCLA in 1969 and had already formed a plan to study the orangutans of Indonesia. After his lecture, they had a long conversation, and Leakey was soon convinced she could undertake the fieldwork in southeast Asia to study the third of our three great ape relatives. By 1971 she was in the jungles of Borneo, and her work laid the foundation for all future studies of orangutan behavior. These three women were pioneers for what is now a large research field in primate behavior. They were unusual in that they were all women (at a time when few women were allowed to do rigorous field research anywhere), and they set the stage for many prominent primate researchers to follow (many of them women, inspired by the Trimates).

Goodall's research on the chimpanzees has had the most impact because she was working with our closest living relative, and she was one of the first to spend months at a time observing her subjects from a nearby vantage point after they had become used to her. Unlike the usual academic method of numbering their field subjects, Goodall gave each chimp a name and soon saw that they had distinct personalities, something no one had observed before. As she wrote, "it isn't only human beings who have personality, who are capable of rational thought [and] emotions like joy and sorrow." She saw so many other human gestures: hugs, kisses, pats on the back, and tickling. These establish "the close, supportive, affectionate bonds that develop between family members and other individuals within a community, which can persist throughout a life span of more than 50 years." The more Goodall watched them, the more they exhibited nearly all the emotional responses that were thought to be unique to humans.

Over the years of observations at Gombe Reserve, Goodall witnessed many behaviors that changed our notions about chimps—and humans. The long-standing belief was that chimps were strict gentle vegetarians, but Goodall saw that they could be aggressive and war-like, hunting colobus monkeys in the trees, killing them, and sharing the carcass among the entire chimp troop. In fact, they killed about a third of all the colobus population at Gombe each year. Even more surprising was the observation of war-like behavior between hostile bands of chimps, or within their troop. She observed dominant females killing the young of other females to maintain their dominance, and sometimes even resorting to cannibalism. As she wrote, "During the first ten years of the study I had believed [. . .] that the Gombe chimpanzees were, for the most part, rather nicer than human beings. [. . .] Then suddenly we found that chimpanzees could be brutal— that they, like us, had a darker side to their nature."

But by far the most groundbreaking discovery was the observation that chimps knew how to make and use tools. Chimps could modify a twig to form a "fishing rod," which they would stick into a termite nest, licking off the termites when they pulled the stick out (figure 23.2). Prior to that time, people often said the line of demarcation between humans and other animals was making tools: "Man the Toolmaker." But chimps are perfectly good toolmakers, even if the tool is simple and not used very often. As Louis Leakey wrote after this discovery, "We must now redefine man, redefine tool, or accept chimpanzees as human!"

Since these discoveries, lots of animals, including gorillas and orangutans, capuchin monkeys, baboons, and mandrills, have been discovered making tools of various kinds for a wide variety of functions, including hunting food (mostly invertebrates and fish), collecting honey, and processing fruits, nuts, vegetables, and seeds. Some tools are used to collect water or to provide shelter, and yes, some primates use weapons during combat that they have fashioned out of local objects. Warfare with weapons is not unique to humans. Both in the wild and in captivity, chimps and other apes have the ability to solve complex problems such as stacking up crates

Figure 23.2 ▲

A chimp using a tool, a modified twig, to fish for termites. (Courtesy of Wikimedia Commons)

to reach bananas high in the cage or fashioning sticks into tools to retrieve bananas beyond the cage bars. They are not doing this by trial and error but by insight, planning, and problem solving—all skills that were once thought to be unique to humans.

The discoveries about primates in the field have been complemented by research on captive chimps, who have demonstrated incredible skills and intelligence never before documented. The most famous of these was a female chimp named Washoe (figure 23.3), who was captured in Africa for Army experiments but then raised almost like a human child by Allen and Beatrix Gardner of the University of Nevada, Reno. Previous studies with other chimps had failed to help them communicate or develop the power of speech, primarily because their voice box isn't shaped the right way for human speech. But chimps have dexterous hands that can form shapes and symbols, so the Gardners tried to teach Washoe to use American Sign Language. They raised Washoe in her own trailer, as if she were a human child, with her own

Figure 23.3 ▲

Washoe, the chimp who learned sign language and broke the communication barrier between apes and humans. (Courtesy of Wikimedia Commons)

couch, bed, refrigerator, clothes, combs, toys, books, and a toothbrush. She underwent a regular child's routine of playtime, chores, riding in the car, and interacting with the Gardners as if they were her parents. Eventually, Washoe learned the signs for more than 350 words and was able to indicate what she wanted or what she thought. When she combined the signs for "water" and "bird" for a swan, she was the first chimp to demonstrate a more abstract kind of reasoning. Harvard psychologist Roger Brown said that her more advanced communication skills were "like getting an SOS from outer space." Eventually, Washoe was capable of signing simple sentences with 5 to 10 words, all of which conveyed complex thought and emotions. Washoe's mate Moja was trained the same way and signed "metal cup drink" for a thermos. After five years, the Gardeners moved on and Roger and Deborah Fouts took care of Washoe at the University of Oklahoma's Institute for Primate Studies. There they observed more signs of human emotions in Washoe. For example, when a caretaker was pregnant and went on maternity leave before miscarrying her child, Roger Fouts wrote:

> People who should be there for her and aren't are often given the cold shoulder—her way of informing them that she's miffed at them. Washoe greeted Kat [the caretaker] in just this way when she finally returned to work with the chimps. Kat made her apologies to Washoe, then decided to tell her the truth, signing "MY BABY DIED." Washoe stared at her, then looked down. She finally peered into Kat's eyes again and carefully signed "CRY," touching her cheek and drawing her finger down the path a tear would make on a human (Chimpanzees don't shed tears). Kat later remarked that one sign told her more about Washoe and her mental capabilities than all her longer, grammatically perfect sentences.

Washoe also showed signs of self-awareness. Seeing herself in the mirror and asked what she saw, she signed "Me Washoe." Jane Goodall argued that this is pretty strong proof that Washoe is aware of herself in the human sense. Washoe was also somewhat confused because she was raised as human, and she behaved toward other chimps as if she thought she were human as well. When students were brought in to "talk" with Washoe, she slowed down her signing so that they could keep up with her.

The achievements of chimps like Washoe are not unique. Another famous "talking" primate was Koko the gorilla. Born in the San Francisco Zoo, a life-threatening illness at age one took her away from her parents to the zoo's hospital. There she was put under the care of Francine Patterson of Stanford University, who worked with her at the Gorilla Foundation compound in Woodside, California, along with male gorillas (who also learned sign language). Koko developed an active vocabulary of more than 1,000 words in her own modified form of "gorilla sign language" (GSL), and she was able to combine signs in new ways to convey new concepts. For example, she had no word for "ring," so she combined the words "finger" and "bracelet" to describe a ring as a "finger-bracelet." She was also taught to listen to and recognize spoken words in English, and Koko eventually learned to recognize about 2,000 words in addition to those she could sign. She used displacement (the ability to communicate about objects that are not present at the moment), recognized herself in the mirror, and reported personal memories. She was also able to use metalanguage to talk about language, such as communicating "good sign" to another gorilla who is doing a good job with sign language. Koko also had her own pet kitten she named "All Ball," and she cared for it as if it were a baby gorilla. When All Ball escaped the cage and was hit and killed by a car, Patterson told Koko, who signed "Bad, sad, bad" and then "Frown, cry, frown, sad, trouble." Koko also made a sound similar to human weeping.

Koko probably didn't use syntax or grammar as it is generally understood in human speech, but her language skills were at least as good as those of a young human child. Some scientists estimated her IQ between 70 and 90, which is slightly lower than average if you administer the same infant IQ test to a young child—but gorilla's brains develop very differently from those of human children and depend more on motor skills, which are different in developing humans.

Many other studies have been done on chimps, gorillas, and other primates, with a wide variety of results. Some are exceptional learners, like Washoe and Koko, and others don't pick up quite as much human emotion or ability to sign as these primate superstars. Nonetheless, they show us how many emotional and psychological traits they have that were once thought to be uniquely human. An entire research field has grown up around primate language, theory of mind, and cognition, and how much our nonhuman relatives can learn to do that humans (especially human children) can do. The sharp boundary between "hooting apes" and humans is now completely blurred.

As Stephen Jay Gould wrote in his essay, "A Matter of Degree," the differences and distinctions between humans and apes are merely a matter of relative size and development (a matter of degree, not a matter of kind). We keep trying to draw a sharp line between humans and other animals, but we cannot find any discretely human feature (the hippocampus, tool use, language, abstract thought) that still can be considered unique to humans. In his words:

> Chimps and gorillas have long been the battleground of our search for uniqueness; for if we could establish an unambiguous distinction—of kind rather than of degree—between ourselves and our closest relatives, we might

gain the justification long sought for our cosmic arrogance. The battle shifted long ago from a simple debate about evolution: educated people now accept the evolutionary continuity between humans and apes. But we are so tied to our philosophical and religious heritage that we still seek a criterion for strict division between our abilities and those of chimpanzees. For, as the psalmist sang: "What is man, that thou art mindful of him? . . . For thou has made him a little lower than the angels, and hast crowned him with glory and honor." Many criteria have been tried, and, one by one they have failed. The only honest alternative is to admit the strict continuity in kind between ourselves and chimpanzees. And what do we lose thereby? Only an antiquated concept of soul to gain a more humble, even exalting vision of our oneness with nature.

FOR FURTHER READING

De Waal, Frans. *Are We Smart Enough to Know How Smart Animals Are?* New York: Norton, 2016.

——. *The Bonobo and the Atheist: The Search for Humanism Among Primates.* New York: Norton, 2013.

——. *Mama's Last Hug: Animals Emotions and What They Teach Us About Ourselves.* New York: Norton, 2019.

——, ed. *Tree of Origin: What Primate Behavior Can Tell Us About Human Evolution.* Cambridge, Mass.: Harvard University Press, 2001.

Fossey, Dian. *Gorillas in the Mist.* New York: Houghton Mifflin, 1983.

Fouts, Roger, with Stephen Tuket Mills. *Next of Kin: What Chimpanzees Have Taught Me About Who We Are.* New York: William Morrow, 1997.

Galdikas, Biruté M. F. *Orangutan Odyssey.* New York: Harry Abrams, 1999.

——. *Reflections on Eden: My Years with the Orangutans of Borneo.* New York: Little Brown, 1995.

Gardner, R. Allen, Beatrix T. Gardner, and Thomas E. Van Cantfort, eds. *Teaching Sign Language to Chimpanzees.* Albany: State University of New York Press, 1989.

Goodall, Jane. *The Chimpanzees of Gombe: Patterns of Behavior.* Cambridge, Mass.: Belknap Press of the Harvard University Press, 1986.

——. *40 Years at Gombe.* New York: Stewart, Tabori, and Chang, 1999.

——. *In the Shadow of Man.* Boston, Mass.: Houghton Mifflin, 1971.

——. *My Friends the Wild Chimpanzees.* Washington, D.C.: National Geographic, 1967.

———. *Through a Window: My Thirty Years with the Chimpanzees of Gombe.* Boston, Mass.: Houghton Mifflin, 1990.

Gould, Stephen Jay. "A Matter of Degree." In *Ever Since Darwin* (1979; repr. New York: Norton, 2007), 49–55.

King, Glenn E. *Primate Behavior and Human Origins.* London: Routledge, 2015.

Montgomery, Sy. *Walking with Great Apes: Jane Goodall, Dian Fossey, Biruté Galdikas.* New York: Houghton Mifflin, 1991.

Morell, Virginia. *Ancestral Passions: The Leakey Family and the Quest for Humankind's Beginnings.* New York: Simon & Schuster, 1995.

Mowat, Farley. *Woman in the Mists: The Story of Dian Fossey and the Mountain Gorillas of Africa.* New York: Warner, 1987.

Patterson, Francine, and Eugene Linden. *The Education of Koko.* New York: Holt Rinehart & Winston, 1981.

Shipman, Pat. *The Animal Connection: A New Perspective on What Makes Us Human.* New York: Norton, 2011.

Silvey, Anita. *Undaunted: The Wild Life of Biruté Mary Galdikas and Her Quest to Save Orangutans.* Washington, D.C.: National Geographic, 2019.

BONES OF OUR ANCESTORS

We're all one dysfunctional family
No matter where we nomads roam
Rift Valley Drifters, drifting home genome by genome
Take a look inside your genes, pardner, then you'll see
We've all got a birth certificate from Kenya

—ROY ZIMMERMAN, "RIFT VALLEY DRIFTERS"

We humans are curious about our origins and our place in nature. Who are we? Where did we come from? What does it mean to be human? Many cultures have creation myths that explain how people fit in with nature and their deity. But as educated people of the twenty-first century, we are interested in the scientific evidence for how humans evolved, and where we came from.

The evidence for human origins has increased exponentially in the past 50 years, and the story is now very well-known (figure 24.1). But it was not always so. At one time, little was known about the fossil record of humans, and the evidence from molecular biology has emerged only in the past 40 years. But the bigger problem is that what most people think they "know" about human evolution is just plain wrong and is based on outdated or distorted or false notions. These "myths" get in the way of any meaningful understanding or having a conversation on this topic. However, just because people have false, outdated, or mistaken notions in their head doesn't mean that the mountains of evidence for human evolution are invalid.

Figure 24.1 ▲

This museum display of replicas of the skulls of the many fossil humans brings home the point that there is now an excellent fossil record of human evolution, with many transitional fossils, from those that are very ape-like to those that are only slightly different from modern humans. (Photograph by the author)

As discussed in chapter 7, *evolution is a bush, not a ladder*. The old idea of a "ladder of life" (*scala naturae*) placed "lower forms" of life such as sponges and corals at the bottom, then molluscs, then fish, amphibians, reptiles, birds, mammals, and us at the top of the ladder. This false notion dates back to the days of Aristotle more than 2,500 years ago, but it became obsolete in 1859 when Darwin showed that nature produces a bushy, branching history—a family tree. Organisms have branched off the family tree of life at different times in the geological past; some have survived quite well as simple corals or sponges, and others have evolved more sophisticated ways of living. Corals and sponges, although simple compared to other organisms, are not "lower" organisms, nor are they evolutionary failures that did not advance up the ladder. They are not trying to evolve into a "higher" organism like a worm or a mollusc. They are good at doing what they do (and have been doing for more than 500 million years), and they exploit their own niches in nature without any reason to change into something different.

The ladder of life was often called the "great chain of being" by pre-Darwinian scholars, and they proposed the ridiculous, now outdated idea of a "missing link" in this chain. The entire concept of a missing link is biologically meaningless, and no scientist uses this term. It is only used by people who don't understand how evolution works. The concept of missing links in human evolution is archaic and useless. We now have thousands of fossils of humans, and these fossils provide a nearly complete record, with only small gaps, which is more than sufficient to demonstrate that humans have evolved.

Nevertheless, this antiquated and long-rejected view of life as a ladder of creation or a great chain of being continues to lurk behind many misunderstandings about biology and evolution. For example, it is still common to hear this question: "If humans evolved from apes, why are apes still around?" The first time scientists hear this question, they are puzzled because it makes no sense whatsoever—until they realize that this person believes in concepts that were abandoned by scientists more than 160 years ago. We now know that nature produces a series of relationships in which lineages branch and speciate, forming a bushy pattern, with ancestral lineages sometimes living alongside their descendants (see figure 7.1). Humans and apes had a common ancestor about 7 million years ago (based on evidence from both fossils and molecular biology), and both lineages have persisted ever since then. Saying "if humans evolved from apes, why are apes still around" is comparable to saying, "if you are descended from your father, why didn't your father die when you were born? Why didn't your grandfather die when your dad was born?" We all understand that children branch off from their parents, overlapping with them in time, but our parents do not automatically die when we are born. Similarly, the human family tree branched off from our common ancestors with the rest of the living apes about 7 million years ago, but both branches are still here.

Likewise, the tendency to put things into simple linear order is a common metaphor for evolution—and also one of its greatest misrepresentations. The iconic image is the classic "monkey-to-ape-to-man march of progress" sequence of organisms walking up the evolutionary ladder (see figure 12.2). This icon of evolution is so familiar that it is parodied endlessly in political cartoons and advertisements. Most people think this is an accurate representation of human evolution, but it is not. *Evolution produces a bushy branching tree of life, not a straight simple ladder!*

In the early days of anthropology, scientists were convinced that the most important thing that made us human was our unusually large brain, which enabled us to do complex tasks, form societies, and use language.

Because it was thought this was what made humans special, anthropologists assumed that our larger brain had appeared earlier than any other feature, including our bipedal posture and our ability to walk and run, unlike most mammals (including chimps and gorillas) that walked on all fours. This dogma was so deeply entrenched that anthropologists rejected fossils that showed humans were bipedal very early, long before they had large brains.

The first truly ancient fossil of humankind to be found was the famous "Taung child" skull of *Australopithecus africanus*, discovered by Raymond Dart in South Africa in 1924 (figure 24.2). The hole for the spinal

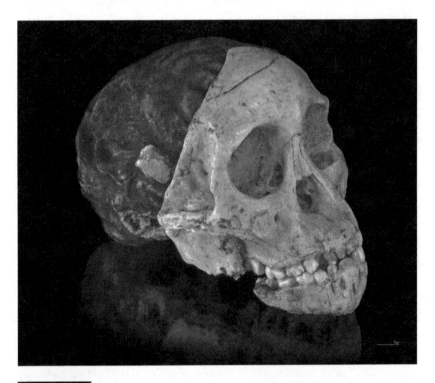

Figure 24.2 ▲

The skull of the Taung child from South Africa, described by Raymond Dart in 1924. The specimen is a juvenile of *Australopithecus africanus*, with a complete face and a natural cast of its brain as well. It was the first important early hominin fossil found outside Eurasia, and it challenged the prevailing notion that humans evolved in Eurasia, not Africa. Most European scientists rejected it at first because it had a small brain yet was clearly bipedal. But decades of further discoveries in Africa proved Dart was right and the others were wrong—most of early human evolution took place in Africa. (Courtesy of Wikimedia Commons)

column (foramen magnum) was directly beneath the center of the skull, clear evidence that it once walked on two feet with an upright posture. Nevertheless, it had a small brain. European scholars rejected this fossil as a human relative for decades because they felt that the brain must enlarge before humans became bipedal or developed any other advanced features. There was also a racist bias among European scholars that the first humans had arisen in Eurasia, not in the Dark Continent of Africa, where people have black skins. In contrast, Darwin thought humans arose in Africa for the simple reason that our closest relatives, chimps and gorillas, live there.

They were also misled by "Piltdown Man," a hoax first announced in 1912 that was put together from the skull of a modern human and the jaw of an orangutan cleverly broken in the right places and stained to make them look ancient. The forger (amateur archeologist Charles Dawson and possibly some accomplices) knew exactly what British anthropologists were expecting, so he used a large-brained medieval human skull and a modern orangutan jaw to make it seem plausible to anthropologists of that time. It was the pride of the British anthropological establishment for years, often proudly described as "the first Briton." The Piltdown forgery was exposed in 1953 as more and more discoveries from Africa showed that humans first evolved there, so the Piltdown "fossil" no longer made sense. When scientists analyzed it carefully, they found evidence of the forgery.

Since the 1950s, all of the most ancient fossils of human relatives have been discovered in Africa, and all of them are bipedal, as far back as the oldest specimens dated 6 to 7 million years ago. Bipedalism was one of the first things to evolve in our ancestry. Meanwhile, our large brain capacity only appeared about 100,000–300,000 years ago at the earliest, so large brains were a very late feature in human evolution.

Probably in part because of the earlier false notion of a linear "march of progress" through time, for decades anthropologists were convinced that only one species of human lived on the planet at any given time. They looked at how *Homo sapiens* has spread throughout the world as a single species today, and they could not imagine that our ancestors could have tolerated another species of human in the same place and time. Thus the linear "march of progress" myth supported the idea that only one species of humans existed at any given time. Paleoanthropologists went to extraordinary lengths to shoehorn all the fossils they were finding into a single species, but by the 1960s and 1970s so many well-preserved fossils from rocks in southern and eastern Africa clearly looked like different species that this idea no longer held up.

As an example of their misguided efforts, they pointed to the striking differences between male and female gorilla skulls and argued that the robust and gracile fossils of australopithecines were just males and females of the same species. Eventually, the presence of multiple different types of skulls and jaws (not just two, as would be expected in males and females) from the same beds, all of the same age and from the same place, made it impossible to hold this notion any more. In addition, paleoanthropologists were coming to realize that evolution is a branching, bushy process, and simplistic linear ancestor-descendant sequences of fossils do not represent the complexity of the real story. Dozens of species of human relatives (hominins, member of our subfamily Hominini) have been found in the fossil record, and sometimes as many as four or five of them lived in the same place. The fossil data are so rich now that this can no longer be denied.

Although most people don't realize it, the fossil record of humans is no longer as poor as it was even 40 years ago. Decades of hard work in the field by hundreds of scientists have turned up thousands of hominin fossils (see figure 24.1), including a few fairly complete skeletons and many well-preserved skulls that clearly show how humans have evolved over 7 million years. This avalanche of new discoveries year after year has occurred despite the fact that hominin fossils are delicate and rare, and only one or two are found for every hundred or more fossil pigs or fossil horse specimens found in the same beds in eastern Africa. Many museums in Africa, Europe, and Asia now have large collections of our early ancestors, so lots of fossils are available for scientists to examine.

The entire story of human evolution is too long and detailed to be discussed in a single chapter, so I will just cover the highlights. The short version is this: dozens of human species and genera are now known, forming a very bushy family tree that spans almost 7 million years of human evolution, mostly in Africa (figure 24.3). The exact details of how all these fossils should be named or how they are interrelated continue to be refined and debated because many of the specimens are incomplete and because of the amazing pace of new discoveries.

The oldest fossil that truly can be described as a member of our own subfamily was discovered and described about 15 years ago. Nicknamed

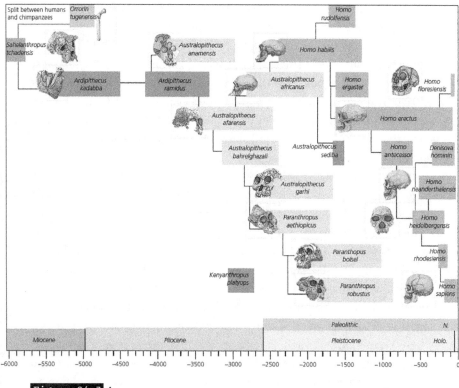

Figure 24.3 ▲

The current family tree of the dozens of different species and genera of fossil humans now known. The exact relationships between the lineages remain controversial, but the existence of all these different species and their time range is well documented.

"Toumai" by its discoverers, its formal scientific name is *Sahelanthropus tchadensis*. The best specimen is a nearly complete skull (figure 24.4A) from rocks about 6–7 million years in age from the sub-Saharan Sahel region of Chad (hence the scientific name, which translates to "Sahel man of Chad"). The skull is very chimp-like with its small size, small brain, and large brow ridges, but it had remarkably human-like features—a flattened face, reduced canine teeth, enlarged cheek teeth with heavy crown wear, and an upright posture—all of this at the very beginning of human evolution. Just slightly younger is *Ororrin tugenensis*, from the upper Miocene Lukeino Formation in the Tugen Hills in Kenya, dated between 5.72 and 5.88 million years ago. *Ororrin* is known mainly from fragmentary remains, but the teeth have the thick enamel typical of early hominins, and the thighbones clearly show that it walked upright. Slightly younger still are the remains of *Ardipithecus kadabba*, found in Ethiopian rocks dated between 5.2 and 5.8 million years ago and consisting of a number of fragmentary fossils. The foot bones show that hominins used the "toe off" manner of upright walking as early as 5.2 million years ago. Thus our human lineage was well established by the latest Miocene and fully upright in posture, even though our brains were still small and our body size not much different than that of contemporary apes.

The Pliocene saw an even greater diversity of hominins (see figure 24.3), with a number of archaic species overlapping in time with the radiation of more advanced hominins. Archaic relics of the Miocene included *Ardipithecus ramidus*, found in Ethiopia in 1992 from rocks 4.4 million years in age, which had human-like reduced canine teeth and a U-shaped lower jaw (instead of the V-shaped lower jaw of the apes). *Ardipithecus ramidus* is now known from nearly complete skeletal material (figure 24.4B), making it the oldest hominin skeleton known. Rocks in Kenya about 3.5 million years in age also yield other more primitive forms, including *Kenyapithecus platyops*.

By 4.2 million years ago, the first members of the advanced genus *Australopithecus*, the most diverse genus of our family in the Pliocene, are also found. The oldest of these fossils is *Australopithecus anamensis* from rocks near Lake Turkana in Kenya ranging from 3.9 to 4.2 million years in age. These creatures were fully bipedal, as shown not only by their bones but also by hominin trackways near Laetoli, Tanzania. In 2019, a complete skull of this species was reported for the first time, which made its anatomy and relationships much better understood. The most famous of these early australopithecines is *A. afarensis* (from rocks 2.95 to 3.85 million years of age near Hadar, Ethiopia), better known as "Lucy" by its discoverer, Don Johanson. Celebrating by the campfire the night after they had just made the discovery, Johanson's team sang along with a tape of the Beatles' song "Lucy in the Sky with Diamonds." Team member Pam Alderman then suggested nicknaming the fossil Lucy (figure 24.4C; figure 24.4D). When it was discovered in 1974, *Australopithecus afarensis* was the first early hominin to clearly show a bipedal posture (based on the knee joint and pelvic bones) but was not as upright as later hominins. These were still small creatures (about 1 meter [3 feet] tall) with small brains, and very ape-like in having large canine teeth and a large protruding jaw.

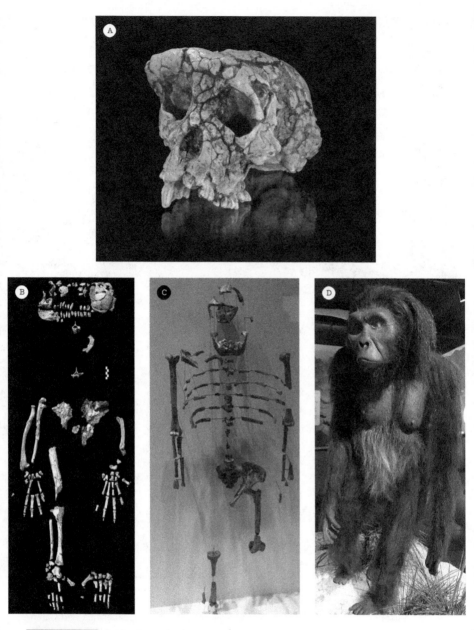

Figure 24.4 ▲ ▶

Some of the best fossils of early fossil hominins. (A) The skull of *Sahelanthropus tchadensis*, nicknamed "Toumai," the earliest known member of our lineage, from beds 6–7 million years in age in Chad. (B) The nearly complete skeleton of *Ardipithecus ramidus*, the oldest hominin fossil known from such a complete skeleton, found in rocks 4.4 million years in age in Ethiopia. (C) The partial skeleton of *Australopithecus afarensis*, nicknamed "Lucy" by its discoverer Don Johanson, from beds about 2.95–3.85 million years in age in Ethiopia. (D) Reconstruction of "Lucy," who was only about 1 meter (3 feet) tall but fully bipedal. ([A] Courtesy of Wikimedia Commons; [B] courtesy of Tim White; [C, D] photographs by the author)

Figure 24.4 ▲ (continued)

(E) Museum display showing the comparison between australopithecines and paranthropines. At the top are the gracile skulls of *Australopithecus africanus*. In the middle row are skulls of three species of *Paranthropus*, showing how much more robust and heavily built they are compared to australopithecines. On the left is the "Black Skull" from beds 2.5 million years in age on the shores of Lake Turkana, the oldest member of the genus *Paranthropus*, *P. aethiopicus*. In the center is a replica of Leakey's famous specimen of *Paranthropus boisei*, called the "Nutcracker Man" based on its powerful jaws and huge molars with thick enamel (see the jaws in the bottom row). It was found in beds 1.9–2.3 million years in age in Olduvai Gorge, Tanzania. On the right is the first named species of the genus, *Paranthropus robustus*. On the bottom row are the lower jaws of paranthropines, showing their enormous broad molars adapted for crushing food, very different from australopithecines. ([E] Courtesy of Bridget McGann)

By the late Pliocene, hominins had become very diverse in Africa (see figure 24.3). These included not only the primitive forms *Australopithecus garhi* (dated at 2.5 million years) and *A. bahrelghazali* (dated at 3.5 million years) but one of the best-known australopithecines, *Australopithecus africanus* (see figure 24.2). Originally described by Raymond Dart in 1924, for decades the Eurocentric anthropology community refused to accept it as ancestral to humans. But as more South African caves yielded better specimens to paleontologist Robert Broom (especially the adult skull nicknamed "Mrs. Ples") and others, it became clear that *A. africanus* was a bipedal, small-brained African hominin, not an ape. *A. africanus* was a rather small, gracile creature, with a dainty jaw, small cheek teeth, no skull crest, and a brain only 450 cubic centimeters in volume. On the basis of its gracile and very human-like features, *A. africanus* is often considered the best candidate for ancestry of our own genus *Homo*.

In addition to *A. africanus*, the late Pliocene of Africa yielded a number of highly robust hominins. For a long time, they were lumped into a very broad concept of the genus *Australopithecus*, either as distinct species or even dismissed as robust males of *A. africanus*. In recent years, however, anthropologists have come to regard them as a separate robust lineage, now placed in the genus *Paranthropus*. The oldest of these is the curious "Black Skull" (so-called because of the black color of the bone), discovered in 1975 on the shores of West Lake Turkana, Kenya, from rocks about 2.5 million years in age (figure 24.4E) and described by my friend, the late Alan Walker. Although it is small in brain size, the skull is robust with a large bony ridge along the top midline of the skull (called a sagittal crest), massive molars, and a concave dish-shaped face. Currently, scientific opinion places the Black Skull as the earliest member of *Paranthropus*, *P. aethiopicus*. It was followed by the most robust of all hominins, *P. boisei*, from rocks in East Africa ranging from 2.3 to 1.2 million years in age (figure 24.4E). The first specimen found of this species was nicknamed "Nutcracker Man" for its huge thick-enameled molars, robust jaws, wide flaring cheekbones, and strong crest on the top of its head, suggesting a diet of nuts or seeds or even bone cracking. Discovered by Mary Leakey at Olduvai Gorge in 1959, it was originally named "*Zinjanthropus boisei*" by Louis Leakey, who made his reputation from it. The rocks of South Africa between 1.6 and 1.9 million years in age yielded the type species of *Paranthropus*, *P. robustus* (figure 24.4E). These too had massive jaws, large molars, and large skull crests but were not as robust as *P. boisei*. *Paranthropus robustus* lived side by side in the same South African caves as *A. africanus*. It is not only more robust but also larger than that species as well, with some individuals weighing as much as 120 pounds.

Finally, the early Pleistocene produces the first fossils of our own genus *Homo*, which are easily distinguished from contemporary *Australopithecus* and *Paranthropus* by a larger brain size, flatter face, no skull crest, reduced brow ridges, smaller cheek teeth, and reduced canine teeth. The first of these to be described was *Homo habilis* (whose name literally means "handy man"), discovered in the 1960s by Louis and Mary Leakey in Olduvai Gorge, Tanzania, from beds about 1.75 million years in age (figure 24.5A).

Originally, all of the earliest *Homo* specimens were shoehorned into the species *H. habilis*, but now paleoanthropologists recognize that this material is too diverse to belong to one species, and several are now recognized. These include the more advanced-looking skull (figure 24.5A) now known as *H. rudolfensis* (dated to about 1.9 million years in age), which made Richard Leakey's reputation, and the very advanced but short-lived *Homo ergaster* (figure 24.5B), from beds 1.6 to 1.8 million years in age. These species are known not only from bones but also from their primitive stone tools, especially choppers and hand axes of the "Oldowan" technology.

Many of the archaic Pliocene taxa persisted into the early Pleistocene (as recently as 1.6 million years ago), including *Paranthropus robustus* and *P. boisei*, *Homo ergaster*, and *Homo habilis*. The best-known fossil of *H. ergaster* is a nearly complete skeleton of a boy who died when he was 8 or 9 years old, found on the shores of West Lake Turkana by Alan Walker and his crew in 1984. Nicknamed "Nariokotome Boy" (figure 24.5B), it is estimated that he would have been 2 meters (6 feet 5 inches) tall if fully grown.

By 1.9 million years ago, however, a new species had appeared: *Homo erectus* (figure 24.6). This human was not only bipedal and stood erect (as its species name implies) but was also almost as large in body size as we are. Its brain capacity was about 1 liter (1,000 cubic centimeters), only slightly less than ours. *H. erectus* made crude choppers and hand axes ("Acheulean culture" tools) and was the first species to make and use fire. Originally, *H. erectus* was confined to Africa, where all of our other ancestors had long lived. But around 1.8 million years ago, we have evidence that *H. erectus* migrated outside our African homeland, as specimens from Indonesia (originally described as "*Pithecanthropus erectus*" or "Java man") have

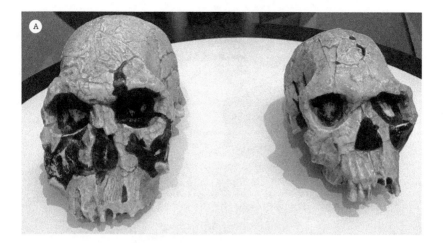

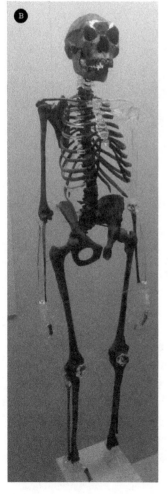

Figure 24.5 ▲

Fossils of the earliest members of the genus *Homo*.
(*A*) Side-by-side comparison of the skulls of *Homo
rudolfensis* (*left*) and *Homo habilis* (*right*), the earliest
species in our genus. (*B*) The nearly complete skeleton of *Homo ergaster*, known as the "Nariokotome
boy," found by Alan Walker and crew on the shores
of West Turkana in 1984. It dates to about 1.7 million
years in age. (Photographs by the author)

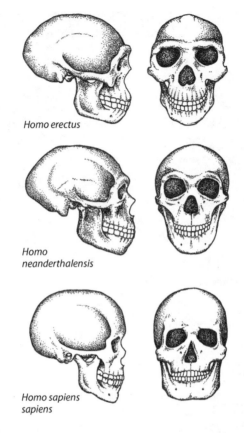

Homo erectus

Homo neanderthalensis

Homo sapiens sapiens

Figure 24.6 ▲

Comparisons of the skulls of *Homo erectus* (*top*), *Homo neanderthalensis* (*center*), and modern *Homo sapiens* (*bottom*). All have relatively large brains, with about 1,000 cubic centimeter capacity in *Homo erectus*, and both Neanderthals and modern humans have brains in the 1,500–1,700 cubic centimeter range. The two extinct species have strong brow ridges, a protruding snout without much of a chin, and broader, heavier cheek bones compared to modern humans. Neanderthals have about the same brain size as modern humans, but their skull is a bit flatter with a point on the back end. (Drawing by Carl Buell)

been dated at that age. In addition, specimens almost as old are known from elsewhere in Eurasia, such as Romania and the Republic of Georgia. By about 500,000 years ago, we have abundant fossils of *H. erectus* in many parts of Eurasia, including the famous specimens from the Chinese caves at Zhoukoudian, originally called "Peking Man" and dated as old as 460,000 years ago. The latest dating suggests that *H. erectus* may have persisted as recently as 74,000 years ago, overlapping with modern *H. sapiens*.

Homo erectus was not only the first widespread hominin species but also one of the most successful and long-lived, spanning more than 1.8 million years in duration between 1.9 and 0.073 million years ago. During much of that long time, it was the only species of *Homo* on the planet and changed very little in brain size or body proportions. If longevity is a measure of success, then it could be argued that *H. erectus* was even more successful than we are.

By about 400,000 years ago, another species was established in western Europe and the Near East: the Neanderthals (see figure 24.3; figure 24.6). In 1857, these were the first fossil humans ever discovered, although their fragmentary fossils were originally dismissed as the remains of diseased Cossacks that had died in caves. The first complete descriptions of skeletons were based on an early specimen from a cave at La Chapelle aux Saints in France that suffered from old age and rickets, so for decades Neanderthals were thought to be stoop-shouldered, bow-legged, and primitive, the classic stereotypical grunting "cave men."

Modern research has shown that Neanderthals were very different from this outdated image. Although their skulls are distinct from ours in having a protruding face, large brow ridges, no chin, and a lower flatter skull that sticks out in the back (see figure 24.6; figure 24.7), they had, on average, a slightly *larger* brain capacity than we do, and they practiced a complex culture that included ceremonial burials, suggesting religious beliefs. Their bones (and presumably bodies) were robust and muscular and slightly shorter than the average modern human. They lived exclusively in the cold climates of the glacial margins of Europe and the Middle East, where their stocky build (like a modern Inuit or Laplander) would have been an advantage. Their tool kits and culture were also more complex, with Mousterian hand axes, spear points, and other complex devices, as well as bone and wooden tools. Some of these tools show complex working and simple carving, so they were artistic as no hominin before had ever been. The famous discoveries at Shanidar Cave in Iraq showed that Neanderthals buried their dead with multiple kinds of colorful flowers, suggesting that they may have had at least some kind of religious beliefs and possibly belief in an afterlife.

For decades, anthropologists treated Neanderthals as a subspecies of *Homo sapiens*, but recent work suggests that they were a distinct species. The best fossil evidence of this comes from the Skhul and Qafzeh caves

Figure 24.7 ▲
Reconstructed skeleton and life-sized model of a Neanderthal. (Courtesy of Wikimedia Commons)

on Mt. Carmel in Israel, where layers bearing Neanderthal remains are interbedded and alternate with layers containing early modern humans. In 1997, Neanderthal DNA was sequenced, and they are clearly not *Homo sapiens* but genetically distinct as well. However, all modern humans of non-African descent have a bit of Neanderthal DNA in them, so there must have been some interbreeding between them in Eurasia where they overlapped.

Neanderthals were the only extinct species of human known from DNA sequencing until 2010, when molecular biology shocked the world with the announcement that there was yet another species of human during the last 40,000 years. Digging in Denisova Cave in the Altai Mountains of Siberia near the Mongolian-Chinese border, Russian archeologists found a juvenile finger bone, a toe bone, and a few isolated teeth of a hominin mixed with artifacts including a bracelet. The artifacts gave a radiocarbon date of 41,000 years ago, so the age was well established. But when the molecular biology lab of Svante Pääbo and Johannes Krause at the Max Planck Institute in Leipzig, Germany (who first sequenced Neanderthal DNA), analyzed the mitochondrial DNA of the finger bone, they found it had a unique genetic sequence that was distinct from both Neanderthals and modern humans. The nuclear DNA was also distinct, but suggested that these people were closely related to the Neanderthals. They may also have interacted with modern humans, because they share 3 to 5 percent of their DNA with Melanesians and Australian Aborigines. The mitochondrial DNA data suggest that they branched off from the human lineage about 600,000 years ago and represent a separate "out of Africa" migration distinct from the much earlier (1.8 million years ago) *Homo erectus* exodus, or the much younger (300,000 years ago) emigration of *H. rhodesiensis-H. heidelbergensis* from Africa to Eurasia.

These mysterious people whose DNA was so distinctive are temporarily called the "Denisovans." There are so few fossils that we cannot say much about their physical appearance or anything else other than that they have distinctive DNA that is found in no other human species. In fact, scientists are still reluctant to give the Denisovans a formal scientific name because there is not enough fossil material to describe the anatomy of the species in any normal sense. So the Denisovans are mysterious, showing us that the bones don't tell the whole tale. Numerous other human species may have lived on this planet but haven't left a fossil record.

Almost as surprising as the 2010 discovery of the Denisovans was the 2003 announcement of a dwarfed species of humans found only on the island of Flores in Indonesia (figure 24.8). Found at a site called Liang Bua Cave, their fossils and artifacts are dated between 1 million and 74,000 years ago. The most striking feature of these people is their tiny size, with a fully grown adult only about 1.1 meters (3 feet 7 inches) tall. They have been nicknamed the "hobbits," but these are not modern African pygmies (which are tiny but fully modern humans). This is an entire population of dwarfed people that appear to have been descended from

Figure 24.8 ▲

The preserved bones of the skeleton of a "Hobbit," the dwarfed human species from Flores Island in Indonesia, *Homo floresiensis*. The entire skeleton is not known, but based on the limb proportions, these people were only about 1.1 meters (3 feet 7 inches) tall as adults. (Photograph by the author)

Homo erectus ancestry (or possibly even from *Homo habilis* ancestry) about a million years ago, then became dwarfed. Size reduction is a common effect on oceanic islands, with many types of animals (especially elephants, mammoths, and hippos) undergoing dwarfing on islands ranging from Malta to Crete to Cyprus to Madagascar. The reason for this dwarfing is clear; they are living on the smaller food resource base of an island and haven't access to sufficient nutrition to grow to normal sizes. In addition, they are typically not under pressure from large predators on islands, nor are they competing with large herbivores. Although the interpretation of these fossils is controversial, most anthropologists agree that they were a distinct species, which has been formally named *Homo floresiensis*.

Finally, we find the first fossil skulls and skeletons that look almost indistinguishable from our own species. Some of these—dubbed "archaic *Homo sapiens*" or more formally, *Homo heidelbergensis*—are known from a few deposits in Africa dating as old as 300,000 years. About 90,000 years ago, skulls from sites in Africa (such as from Klasies Mouth Cave in South Africa) look almost completely modern in appearance and are universally regarded as *Homo sapiens* (our species). Like *Homo erectus*, early *H. sapiens* spent most of their history in Africa and migrated to Asia about 200,000 years ago, and then into Europe about 70,000 years ago. There these people came into contact with Neanderthals, and for about 30,000 years they coexisted. Mysteriously, Neanderthals vanished 40,000 years ago. Whether they were wiped out by *H. sapiens* or by some other cause is not clear. The subject has been one of endless debate and speculation. Pat Shipman argues that modern humans had an advantage in domesticated dogs, which helped them overcome Neanderthals in hunting and in warfare. Whatever happened, modern *H. sapiens* soon took over the entire Old World, developing complex cultures (the "Cro-Magnon people") including the famous cave paintings of Europe, and many kinds of weapons and tools.

This brief review of the hominin fossil record hardly does justice to the richness and quality of the specimens or to the incredible amount of anatomical detail that has been deciphered. If it all seems like too much to absorb, just gaze at the faces of the skulls in figure 24.1. They look vaguely like modern human skulls, but they definitely show the change from more primitive hominins that some people see as "mere apes" (even though they were all completely bipedal and had many other human characteristics) up through forms that everyone would agree look much like "modern humans" (even though they had many distinctive anatomical features, like those found in Neanderthals, that make them a distinct species). Even nonscientists can glance at these fossils and see the hallmarks of their own ancestry.

FOR FURTHER READING

Beard, Chris. *The Hunt for the Dawn Monkey: Unearthing the Origin of Monkeys, Apes, and Humans*. Berkeley: University of California Press, 2004.

Conroy, Glenn C. *Primate Evolution*. New York: Norton, 1990.

Delson, Eric, ed.. *Ancestors: The Hard Evidence*. New York: Liss, 1985.

Harari, Yuval Noah. 2015. *Sapiens: A Brief History of Humankind*. New York: Harper, 2015.

Johanson, Donald, and Maitland Edey. *Lucy: The Beginnings of Humankind*. New York: Simon & Schuster, 1981.

Johanson, Donald, and Blake Edgar. *From Lucy to Language*. New York: Simon & Schuster, 1996.

Johanson, Donald, and Kate Wong. *Lucy's Legacy: The Quest for Human Origins*. New York: Crown, 2009.

Johanson, Donald, Lenora Johanson, and Blake Edgar. *Ancestors: In Search of Human Origins*. New York: Villard, 1994.

Jurmain, Robert, Lynn Kilgore, Wenda Trevathan, Russell L. Ciochon, and Eric Bartelink. *Introduction to Physical Anthropology*. 15th ed. New York: Cengage Learning, 2017.

Larsen, Clark Spencer. *Our Origins: Discovering Physical Anthropology*. New York: Norton, 2014.

Lewin, Roger. *Bones of Contention: Controversies in the Search for Human Origins*. Chicago: University of Chicago Press, 1987.

——. *In the Age of Mankind: A Smithsonian Book on Human Evolution*. Washington, D.C.: Smithsonian Institution Press, 1988.

Lewin, Roger, and Robert A. Foley. *Principles of Human Evolution*. 2nd ed. Malden, Mass.: Wiley-Blackwell, 2003.

Pääbo, Svante. *Neanderthal Man: In Search of Lost Genomes*. New York: Basic Books, 2014.

Roberts, Alice. *Evolution: The Human Story*. London: DK Books, 2018.

Sawyer, G. J., Viktor Deak, Esteban Sarmiento, and Richard Milner. *The Last Human: A Guide to the Twenty-Two Species of Extinct Humans*. New Haven, Conn.: Yale University Press, 2007.

Shipman, Pat. *The Invaders: How Humans and Their Dogs Drove Neanderthals to Extinction*. Cambridge, Mass.: Belknap Press of Harvard University Press, 2015.

——. *The Man Who Found the Missing Link: Eugene Dubois and His Lifelong Quest to Prove Darwin Right*. New York: Simon & Schuster, 2001.

Stringer, Chris. *Lone Survivors: How We Came to Be the Only Humans on Earth*. London: Times Books, 2012.

Stringer, Chris, and Peter Andrews. *The Complete World of Human Evolution*. London: Thames & Hudson, 2005.

Stringer, Chris, and Clive Gamble. *In Search of Neanderthals: Solving the Puzzle of Human Origins*. London: Thames & Hudson, 1993.

Swisher, Carl, III, Garniss Curtis, and Roger Lewin. *Java Man*. New York: Scribner, 2000.

Tattersall, Ian. *The Human Odyssey: Four Million Years of Human Evolution*. Upper Saddle River, N.J.: Prentice Hall, 1993.

——. *Masters of the Planet: The Search for Our Human Origins*. London: St. Martin's Griffin, 2012.

——. *The Strange Case of the Rickety Cossack and Other Cautionary Tales from Human Evolution*. New York: St. Martin's Press, 2015.

Tattersall, Ian, and Jeffrey Schwartz. *Extinct Humans*. New York: Westview, 2000.

Walker, Alan, and Pat Shipman. *The Wisdom of the Bones: In Search of Human Origins*. New York: Knopf, 1996.

THE ONCE AND FUTURE HUMAN

I teach you the Übermensch. Man is something that shall be overcome. What have you done to overcome him? . . . All beings so far have created something beyond themselves; and do you want to be the ebb of this great flood, and even go back to the beasts rather than overcome man? What is the ape to man? A laughingstock or established embarrassment. And man shall be that to Übermensch: a laughingstock or painful embarrassment. You have made your way from worm to man, and much in you is still worm. Once you were apes, and even now, too, man is more ape than any ape. . . . The Übermensch is the meaning of the earth. Let your will say: the Übermensch shall be the meaning of the earth. . . . Man is a rope, tied between beast and Übermensch–a rope over an abyss . . . what is great in man is that he is a bridge and not an end.

—FRIEDRICH NIETZSCHE, *THUS SPOKE ZARATHUSTRA* (1883)

Science fiction has filled our imagination with all sorts of images and ideas about how humans might evolve in the future, and what we might look like a hundred or a thousand years from now. Many of these ideas envision humans developing larger and larger brains until we have huge bulbous bald heads, or developing huge eyes. In the meantime, our bodies and limbs would degenerate as we become more and more dependent on technology. In the movie *Wall-E*, the humans had become obese blobs dependent on low gravity and all sorts of technology to function in their spacecraft as they abandoned a devastated Earth. Mike Judge's 2006 movie *Idiocracy* envisioned a world in which humans had gotten progressively stupider and stupider over 500 years of being dumbed down by the media and big corporations. They had reached the point where the president of the United

States is a blowhard ex-wrestler who has sex with porn stars and who blusters and puts on a big show for his voters but doesn't have a clue what he's doing. So two time travelers (played by Luke Wilson and Maya Rudolph) who had just average intelligence when they were put in the time capsule 500 years earlier end up being the smartest people on the planet.

Even more pessimistic was H. G. Wells's *The Time Machine*, when humans had devolved into two races, the orc-like subterranean Morlocks and the elf-like forest-dwelling Eloi. Other dystopian visions of humanity include the "Epsilon-Minus Semi-Morons" of Aldous Huxley's *Brave New World* and "The Marching Morons" of Cyril Kornbluth. But the *Avatar* movie envisioned humans that are 12 feet tall with blue skin and a telepathic ponytail. Marvel Comics imagined mutant humans developing into superheroes with superpowers, such as the X-Men. Friedrich Nietzsche envisioned a "super-human" (*Übermensch*) as the natural product of human evolution. On one website, Dr. Alan Kwan of Washington University (who does computational genomics, not evolutionary biology) predicted that humans would have larger brains and huge eyes.[1] Supposedly, we will need them for the dim light of space travel, and we will have more pigmented skin to alleviate the damaging impact of the higher levels of UV in space. In 2019, Astronomer Royal Martin Rees proclaimed that humans would become a new species when they traveled through space, having large brains and amazing vision.[2]

Many of these ideas have been popular in science fiction and pop media for almost a century, but very little of it was based on understanding how evolution works, or what the fossil record shows about how humans have evolved in the past. A quick review of the human fossil record provides a very different perspective (see chapter 24). For one thing, skeletally modern humans (as defined by the bony features that fossilize) have been around for at least 100,000 years, and possibly 300,000 years. Nearly all the fossil *Homo sapiens*, or our immediate ancestors such as *Homo heidelbergensis*, had (on average) about the same cranial capacity as modern humans. In fact, our close relatives, the Neanderthals, had slightly larger brain capacity on average, but they didn't develop the complex tool kit or other advances of culture found in contemporary *Homo sapiens*.

Not only that, but human brains on average have been *decreasing* slightly in size over the past 20,000 years,[3] possibly due to the slightly smaller body size in humans of the last 10,000 years. Brain size is much more strongly correlated with body size than with intelligence, so women on average have smaller brains than men because of their average smaller body size—but they have the same range of intelligence. Others think that the smaller brain arose about 10,000 years ago as civilization made it less necessary to cope with the complex hostile environment hunter-gatherer societies faced. In addition, larger brains are not always good for us; they are energetically very expensive, and the human body must sacrifice other functions to support them. The science fiction scenarios that imagine bulbous heads for huge brains are not consistent with the fossil record, which shows almost unchanging brain size over 300,000 years or shrinking in the last 10,000 years, even as we get smarter. The science fiction world or pop media does not seem to know about the lack of correlation between brain size and intelligence.

When I took some call-in questions after an interview I did on a radio show about evolution, I got a good dose of the misconceptions about human evolution that remain popular today. Again and again, questioners speculated about how humans might evolve in the future. They repeated the false premise that brains are getting larger and that larger brains meant greater intelligence, and they supported another misunderstanding about evolution as well—that evolution is making us better and better and that eventually we will reach the optimal state of the human body. As we learned in this book, natural selection is not about making perfect organisms, or refining an organism until it reaches perfection (a common false notion in early twentieth-century literature). As discussed in chapters 9 and 21, evolution is about adapting to the local conditions of the time, and lots of vestigial organs and useless features and junk DNA are carried along in the process. This material is seldom lost because it doesn't detract from the fitness of the organism to the local set of conditions at a specific time when breeding occurs and the genes are passed to a new generation—and the cost of getting rid of these features, or rewiring the genome to get rid of them, is too high. Pop science and science fiction are full of false premises and outdated notions that ignore these realities.

The realization that the human skull and skeleton had changed very little since about 100,000 years ago or earlier became widely accepted in the late 1960s through the 1980s. Consequently, many leading evolutionary biologists, including Stephen Jay Gould, Ernst Mayr, Richard Dawkins, and David Attenborough, argued that humans had completely stopped evolving in their physical skeleton and in most parts of their anatomy. Because it takes hundreds of generations to make a change in the phenotype, the prevailing idea was that rapid changes in culture made phenotypic and genetic evolution much less necessary. We now adapt by substituting tools and weapons, domesticating animals, and eventually using science and technology, and none of these things required more than a small change in our physical makeup. Culture is indeed capable of extremely rapid changes in years or weeks or days, and in the age of the internet ideas and discoveries are propagated around the world in seconds, whereas physical changes and changing in our genome take many dozens of generations and hundreds to thousands of years. In fact, Richard Dawkins originally coined the word "meme" to describe a cultural feature that is spread rapidly through communication and copying, in contrast to the "gene," which only spreads and changes slowly in response to generations of breeding. Now the word "meme" has been appropriated for almost any idea, and especially for provocative or memorable ideas and graphics that are spread over the internet.

But the idea that humans have stopped physically evolving was too simplistic and has since been proven wrong. The evidence shows that current and future human evolution is not the kind of evolution sci-fi writers envisioned. It's not about humans developing bulging heads with bigger brains, or bigger eyes, or degenerate bodies and limbs. Instead, most of the change is subtle and a lot of it is internal (physiological or genetic) and not obvious to most people, and these changes are not preserved in the bones that might become fossilized.

The first constraint is to realize how recently all the populations of humans on Earth diverged from one another. This was first discovered in the 1990s, when genetic analysis showed that all the peoples of Earth are extremely genetically similar (figure 25.1). All modern humans alive are more similar to each other genetically than populations of chimps from central Africa, eastern Africa, and western Africa are to each other. This could only happen if all the humans on Earth were reduced to a very small number (in the order of 1,000 to 10,000 breeding pairs) and we are all their descendants, with only the limited number of genes that this small number of ancestors originally carried. This phenomenon is known as a *genetic bottleneck*. The genetic diversity of the surviving population is often very low, with an unusually high frequency of some peculiar genes. This is easily detectable by doing a genetic analysis and using molecular clock techniques to calculate how long ago this population crash must have occurred. Many animals alive today went through some kind of bottleneck when their populations were near extinction level. Cheetahs, for example, went through

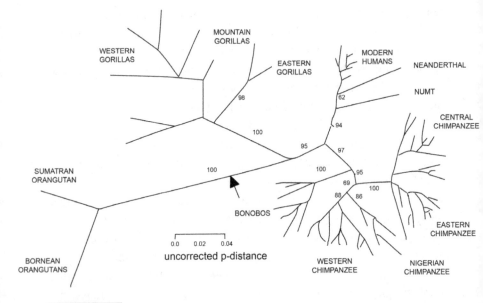

WESTERN
GORILLAS

MOUNTAIN
GORILLAS

EASTERN
GORILLAS

MODERN
HUMANS

NEANDERTHAL

NUMT

CENTRAL
CHIMPANZEE

98

100

95

62

94

97

100

95

69

88

86

100

SUMATRAN
ORANGUTAN

100

BONOBOS

EASTERN
CHIMPANZEE

0.0   0.02   0.04

uncorrected p-distance

BORNEAN
ORANGUTANS

WESTERN
CHIMPANZEE

NIGERIAN
CHIMPANZEE

Figure 25.1 ▲

Molecular phylogeny of apes and humans, showing their genetic distance from one another based on mitochondrial DNA. All human "races" are much more similar to one another than two populations of gorillas or chimpanzees are to each other. (Redrawn from Pascal Gagneux et al., "Mitochondrial Sequences Show Diverse Evolutionary Histories of African Hominoids," *Proceedings of the National Academy of Sciences of the United States of America* 96, no. 9 [1999]: figure 1b; © 1999, used by permission of the National Academy of Sciences USA)

a bottleneck about 10,000 years ago, and today they have a low genetic diversity and are highly inbred, which threatens their survival. The same is true of both the European and American bison; both were hunted nearly to extinction but have since recovered. The northern elephant seal was hunted down to about 30 individuals in the 1890s, but they have now recovered and are threatened by their low genetic diversity.

Bottlenecks seem to have occurred in lots of species, but most especially in humans. It's shocking to think that the earth today is populated by nearly 8 billion people and that number is rapidly increasing but that the human population was much smaller in prehistory. In 2005, a group of scientists found genetic evidence suggesting that only about 70 individuals gave rise to all the Native American populations that spread from Asia to the Americas 18,000 to 11,000 years ago. This means that all native peoples, from the Inuit of Alaska to the Lakota of the Plains to the Incas and

Aztecs and Tolmecs and Mayans and even the Fuegians of Patagonia, are extremely closely related and very low in genetic diversity.

In fact, studies of the full diversity of living *Homo sapiens* show they went through a very small bottleneck not that long ago. Estimates vary, but the most recent genetic evidence places the total modern *H. sapiens* population in the bottleneck down to 30,000 people and maybe as few as 4,500 people, and some estimates are as low as 40 breeding pairs. Most estimates suggest that only about 5,000 people were on Earth at that time. This is less than the population of a small town in America today!

When did humans go through that bottleneck? The latest archeological evidence places it at least 48,000 years ago at a bare minimum. Not many archeological sites have been found during this time interval, so it could be much earlier. We can use the molecular clock to estimate how long ago all the modern humans on the planet diverged from a small survivor population. Some recent genetic studies place it at around 74,000 years ago.

An estimate of 74,000 years ago is an interesting coincidence because that is the date of one of the largest volcanic eruptions in Earth history, the explosion of the Toba volcano in the north end of the island of Sumatra in Indonesia. It was the largest eruption on the planet in the past 28 million years, over a thousand times larger than the catastrophic eruptions of Mount Tambora in Indonesia in 1815 or Krakatoa volcano between Java and Sumatra in Indonesia in 1883, both of which changed the climate for several years after eruption. In 1816, Tambora caused a "year without summer," but by contrast, Toba caused global temperature to drop by 3–5°C (5–9°F), further amplifying the cold of the ongoing Ice Ages. The tree line and snow line dropped 3,000 meters (9,000 feet), making most high elevations uninhabitable. Global average temperatures dropped to only 15°C after three years, and it took a full decade to recover to pre-eruption temperatures. Ice cores from Greenland show the evidence of this dramatic cooling in trapped ash and ancient air bubbles.

What happened to the people and animals during this terrible time? Geneticists and archeologists have found evidence that the Toba catastrophe nearly wiped out the human race. Apparently only a few thousand humans survived worldwide and passed through the genetic bottleneck. Another study found a similar genetic bottleneck in the genes of human lice, and in our gut bacterium *Helicobacter pylori*, which causes human ulcers; both of these date to the time of Toba, according to their molecular clocks,

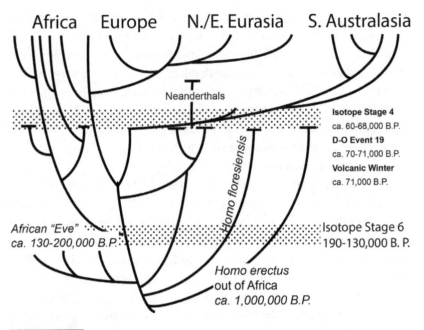

Africa　Europe　N./E. Eurasia　S. Australasia

Neanderthals

Isotope Stage 4
ca. 60-68,000 B.P.
D-O Event 19
ca. 70-71,000 B.P.
Volcanic Winter
ca. 71,000 B.P.

Homo floresiensis

African "Eve"
ca. 130-200,000 B.P.

Isotope Stage 6
190-130,000 B. P.

Homo erectus
out of Africa
ca. 1,000,000 B.P.

Figure 25.2 ▲

There were many different human lineages on the planet before the Toba catastrophe 74,000 years ago, but most of them (including the Flores people and possibly *Homo erectus*) vanished. Only a few (such as the Neanderthals and the ancestors of all modern non-African humans) survived the event and lived on through the latest Pleistocene. (Redrawn from several sources)

which show how long since a genetic change took place. The same is true for the genes of a number of other animals, including tigers, orangutans, several monkeys, gorillas, chimpanzees, and pandas—just about every large mammal in southern Asia or Africa that has had its DNA sequenced so far. In short, Toba was the largest eruption to occur since humans have been on Earth, and it came very close to wiping out humans altogether, along with many other animals. The Toba catastrophe model is still very controversial among anthropologists, although the genetic bottleneck is not in doubt and that it happened around the time of the gigantic Toba eruption is strongly supported by the data.

The genetic evidence for the "out of Africa model" strongly supports the idea that the ancestors of all living people in Eurasia and the Americas did not leave Africa until about 70,000 years ago, some time after the Toba catastrophe (figure 25.2). Apparently, they migrated along the southern coastline of Asia and colonized Australia about 50,000–65,000 years ago. Another lineage of this migration left the African and Middle Eastern region (which had modern humans some time earlier, possibly 300,000 years ago) and reached southern Europe about 55,000 years ago. Finally, the lineages that crossed the Bering land bridge reached the Americas about 13,000–16,000 years ago, although some controversial dates from the Monte Verde site in Chile suggest that some early pioneers might have arrived as early as 30,000 years ago.

Some people may find it disconcerting to learn that all the human "races" are subtle recent variations in human evolution and that they represent less difference in the genome than the differences between populations of chimpanzees in west Africa and those in central Africa. Even more shocking is the idea that the Toba eruption almost wiped humans off the face of the earth 74,000 years ago. Realizing that the differences between all of the so-called races are just tiny, very recent changes in the genome of humanity can certainly change your perspective. In this time of race-baiting in politics and intense hatemongering against immigrants and foreigners, it's important to remember that all humans of any skin color are 99.999 percent identical genetically, and the small changes in skin color and a few other superficial features are very recent and meaningless in scientific and evolutionary terms.

About 30,000 years ago, all humans on Earth lived in the tropics or subtropics, and all of them were very dark-skinned. Most of the skin color differences are adaptations to light levels and differences in solar radiation. Darker pigmentation helps prevent UV radiation from causing sunburn and skin cancer (advantageous to peoples of tropical and subtropical regions), but in regions with much less light (such as the high latitudes of the Northern Hemisphere), light levels are so low that the skin carries very little pigmentation so the solar energy can penetrate the skin and create vitamin D. These tiny differences in skin color by latitude and total amount of solar radiation developed independently in a number of northern Eurasian populations in the past 20,000 years or less (after the last glacial maximum, as humans first returned to subpolar regions when the polar ice caps retreated). Based on genetic evidence, the spread of lighter-skinned populations across Europe is thought to have occurred about 5,000 years ago, during the Mesolithic. Humans are now so mobile and spread so rapidly around the earth that skin color is no longer controlled by local conditions. And humans are interbreeding rapidly and freely so fast that whatever regional differences once existed are largely being blended out as more and more children are born to interracial and mixed-race couples. In addition, artificial means of compensating for skin color differences are available, such as sunscreen to prevent light-skinned people from burning and vitamin D tablets, so no one needs to go without.

Many other small, subtle genetic changes have happened in the last few thousand years. Before 30,000 years ago, all humans had sticky earwax and produced a lot of sweat. About 20,000 years ago, a mutation called ABCC11 arose in mostly eastern Asian populations, and all of their descendants now have dry, flaky earwax and tend to sweat a lot less.

So what physical differences might be recorded in our skeletons and visible to anthropologists of the future? The first thing you'd notice is that about 11,000 years ago, when the first agricultural civilizations arose and people began cooking their food, their teeth and jaws shrank about 10 percent compared to Ice Age peoples who spent their time hunting and gathering. As pointed out in chapter 21, our faces and mouths are getting shorter, with less room for our teeth. The last molars in our jaws (the "wisdom teeth") are often crowded and can become impacted, and in some human populations, they fail to develop at all. Because we no longer need a heavy-duty set of molars to grind hard seeds and nuts, the genes for people without wisdom teeth will spread, and future humans will no longer even develop them.

More recently, the average height of people in Western industrialized nations has increased on average about 10 centimeters (4 inches) in the past 150 years. In 1880, the average American male was only 5 feet 7 inches tall, and now he's 5 feet 10 inches. This increase in human height is almost certainly due to improvements in diet. This was even more dramatically demonstrated in Japan during the 1940s. Before World War II, most Japanese lived in poverty and had poor diets consisting mostly of rice with limited vegetables and protein. After the war was over and Japan modernized during the U.S. occupation, their economy and diet improved, with a lot more protein and healthy fruits and vegetables. As a result, the average height of the postwar generations is noticeably taller than that of their grandparents' generation. This change was so rapid (just a few decades) that it is not coded in the genome yet, but it is an example of ecophenotypic variation—short-term, nongenetic responses to immediate conditions that affect growth and development.

Another example of local physical adaptation is the short-limbed, stocky, fat-retaining build of Inuits (Eskimos) and Laplanders, who live in extremely cold climates. Long limbs are a liability in cold climates because their increased surface area loses heat—a phenomenon known as Allen's rule in biology. This genetic modification also must have begun less than 20,000 years ago, when the glaciers began to first retreat from the polar regions and made them habitable.

The selective advantage of other genes is less obvious. Blue eyes, for example, are due to a gene that appeared about 14,000 years ago in Italy and in the Black Sea region, according to genetic analysis. For some reason, it was considered attractive and selected for, conferring a 5 percent greater chance for successful reproduction, and now it is spread to about half a billion people, mostly of northern Eurasian origins.

Other changes to humans are in their physiology and genetic characteristics rather than their anatomy. One example of recent adaptations is the ability of human populations who live in high mountains (for example, the Sherpas of the Himalayas) to function at lower oxygen levels. That adaptation has arisen independently in humans at least three times.

Before 8,000 years ago, most humans (and even today in cultures with almost no dairy consumption) lost their ability to digest lactose after being weaned from their mother's milk, and they became lactose intolerant. Today many human populations retain the genes that produce lactase for the rest of their lives, so they can digest milk and dairy products and use this source of protein long after they have stopped nursing at their mother's breast. That development appears to have evolved at least 5 times in the past 8,000 years as domesticated animals were kept for milk production. Today about 95 percent of northern European peoples have the gene to make lactase, but it is much less frequent in cultures that don't cultivate animals for milk. In addition, human populations descended from farming cultures produce a lot more copies of the genes that make salivary amylase, which is needed to digest starch; small hunter-gatherer cultures in Africa have far fewer of these genes.

Humans are rapidly developing genes for resistance to various diseases. Sickle cell anemia is widespread in Africa and in populations derived from Africa. When someone gets a copy of the recessive gene (homozygous) from both parents, it is normally fatal. But people who have a copy of both alleles of the gene (heterozygous) have a resistance to malaria. This helps the deadly sickle cell gene persist because heterozygous individuals have an advantage in surviving malaria, whereas people with the homozygous dominant gene are more vulnerable to malaria. As long as the heterozygotes have an advantage, they keep the deadly recessive gene in the population. But another gene, called DARC, that confers resistance to one of the two vectors that carry malaria began appearing in these populations about 45,000 years ago. So the advantage of the sickle cell gene to fight malaria is being lost, and eventually it will disappear. New genes to fight leprosy and tuberculosis are spreading through human populations as well.

We have come a long way from the evolution of the universe and the solar system to the evolution of life on Earth. By following all the clues, both subtle and direct, we now have a template for how life has evolved and continues to evolve. We have seen how closely humans are connected to the rest of life, from our 99 percent similarity in DNA with a chimpanzee and gorilla, to the incredible amount of common behavior and psychology we share with them, to the fossil evidence about how humans evolved—and the data that show we are still evolving. As Charles Darwin first wrote in 1859,

> there is grandeur in this view of life, with its several powers, having been originally breathed into a few forms or into one; and that, whilst this planet has gone cycling on according to the fixed law of gravity, from so simple a beginning endless forms most beautiful and most wonderful have been, and are being, evolved.

Or, as the great geneticist Theodosius Dobzhansky wrote in 1973, "Nothing in biology makes sense except in the light of evolution."

NOTES

1. Cynthia McKanzie, "What Will Humans Look Like in the Future? Some Possible Scenarios," MessageToEagle.com, May 15, 2017, http://www.messagetoeagle .com/will-humans-look-like-future-possible-scenarios/.
2. "Martin Rees: We Will Become a New Species by Expanding Beyond Earth," *New Scientist*, December 18, 2019, https://www.newscientist.com/article /mg24432614-400-martin-rees-we-will-become-a-new-species-by-expanding -beyond-earth/.

3. Cynthia McKanzie, "Why Did Our Brains Start Shrinking 20,000 Years Ago?," MessageToEagle.com, April 29, 2017, http://www.messagetoeagle.com /brains-start-shrinking-20000-years-ago/.

FOR FURTHER READING

Barnett, Adrian. "Future Humans: Just How Far Can Our Evolution Go?" *New Scientist*, no. 3098 (November 2, 2016).

Cochran, Gregory. *The 10,000 Year Explosion: How Civilization Accelerated Human Evolution*. New York: Basic Books, 2009.

Hawks, John. "Still Evolving After All These Years." *Scientific American* 311, no. 3 (2014): 86–91.

Manco, Jean. *Ancestral Journeys: The Peopling of Europe from the First Ventures to the Vikings*. London: Thames & Hudson, 2016.

Prothero, Donald R. *When Humans Nearly Vanished: The Eruption of Toba Volcano*. Washington, D.C.: Smithsonian Books, 2018.

Reich, David. *Who We Are and How We Got Here: DNA and the New Science of the Human Past*. New York: Pantheon, 2018.

Rutherford, Adam. *A Brief History of Everyone Who Ever Lived: The Human Story Retold Through Our Genes*. New York: The Experiment, 2016.

Solomon, Scott. *Future Humans: Inside the Science of Our Continuing Evolution*. New Haven, Conn.: Yale University Press, 2016.

Wade, Nicholas. *Before the Dawn: Rediscovering the Lost History of Our Ancestors*. New York: Penguin, 2006.

——. *A Troublesome Inheritance: Genes, Race, and Human History*. New York: Penguin, 2014.

Wells, Spencer. *Deep Ancestry: Inside the Genographic Project*. Washington, D.C.: National Geographic, 2006.

INDEX

9+2 fiber microstructure, 238

aardvarks, 222
abalones, 267
ABCC11 gene, 340
abdominal pregnancy, 279
abiogenesis, 225
Aborgines, 17, 285
Aboriginal Australian mythology, 17
Acanthostega, 137–146
Acheulean culture, 322
Achilles tendon, 280
Adam (biblical figure), 17, 121
adenine, 242–243
ADP, 236
Adriosaurus, 162–165
"African unicorn," 204
After Many a Summer Dies the Swan (A. Huxley), 295
Age of Dinosaurs, 18–19
Age of Enlightenment, 20
Age of Mammals, 18
Age of Reason, 20
age of the earth, 25–27
agricultural revolution, 340
Ahlberg, Per, 143
AIDS, 247, 254, 293
Aistopods, 166
Akkadians, 72–73
alcohols, 230–233

Ali, Prince Muhammad, 196
Alien (1979 film), 100
Allen's rule, 341
Alpher, Ralph, 12
Alula, 49–50, 172
Amazon rain forest, 74–76
Ambulocetus, 128–130
Ambystoma, 294
amebelodontinae, 217
American Museum of Natural History, 190, 221
amino acid sequencing, 93–94, 245
amino acids, 228–235
ammonia, 228–229
amoebas, 261–262
amphibians, 91, 137–148, 152–155
amphisbaenians, 165
analogy, 49–51, 54
Anchitherium, 182–190
Anderson, Jason, 155–156
Andrews, C. W., 213–215
anglerfish, 112
angular bone, 58–59
angular unconformity, 24–25
anole lizards, 43
antediluvian world, 213
antennapaedia, 255–256
anthracotheres, 127, 134
antibiotics, 238–239
antibodies, 93

antigens, 93
anvil (incus bone), 58–59
aorta, 201–203, 280
apodans, 165
appendix, 283
apples, 40
archaea, 234–235
archaebacteria, 234–235
archaeocetes, 127–130
Archaeohippus, 187–190
Archaeopteryx, 170–173, 177–194, 182
archangels, 151
archetype, 53
Arctic islands, 219
Ardipithecus, 87
Ardipithecus kadabba, 318
Ardipithecus ramidus, 318–320
Arenahippus, 189
Argentina, 156, 165, 176
Aristotle, 47–49, 61, 86, 151, 312
Armagh, Ireland, 17
arsinoitheres, 221
arthropods, 54–57, 265–266
articular bone, 58–59
artiodactyls, 132–134, 182
asses, 191–192
atavism, 176, 192, 284, 293
ATP, 236
Attenborough, David, 334
Augustus (Roman emperor), 208
aurochs, 73
Australia, 16, 43, 73–75
Australopithecus, 87
Australopithecus afarensis, 318
Australopithecus africanus, 314, 321
Australopithecus anamensis, 318
Australopithecus garhi, 321
Aves, 91

baboons, 304
Babylonians, 72–73
back problems, 277
bacteria, 19, 46, 100, 150
Baer, Karl Ernst von, 53, 64–67; and laws of
    embryology, 66
baleen whales, 131–132
bamboo, 113–116

barnacles, 109
Barytherium, 214–215
Bates, William Henry, 79
bats, 48–51, 172
Beadnell, H. J. L., 213–215
Beak of the Finch, The (Weiner), 31–32, 39
Becquerel, Henri, 26
bees, 58
beetles, 55
Bell, Thomas, 80
Bell Labs, 11–13
belly ribs, 172
Belon, Pierre, 50–52
Bendix-Almgreen, Svend, 144
Bentham, Jeremy, 20
Bergen, Norway, 42
Bergmann's rule, 41
beriberi, 292
Berkeley, George, 20
Berlin, 64, 170
"best of all possible worlds," 273–274
bichir, 148
Big Badlands, South Dakota, 182, 189
"Big Bang," 11–13
"Big Bird" (Galápagos finch), 38
Big Bone Lick, Kentucky, 209
big toe, 250
binomen, 87
biogeography, 72–84
bipedal posture, 276–280
birds, 49–53; evolution of, 168–177; teeth of,
    169–170, 176
birth defects, 279
Biston betularia, 44–45
bithorax mutant fly, 255–256
black bears, 126–127
Black Sea, 341
black shales, 234
"Black Skull," 320–321
"black smokers," 234–237
Black, Joseph, 20
blastula, 64
blind cave animals, 122
blind spot, 267–269, 276
blood pressure, 200–201
blowhole, 125
blue eyes, 341

body hair, 282
Boltwood, Bertram, 27
Borneo, 303
Boyle, Robert, 108
Brahmatherium, 203
brain size, 323, 333–334
brains before bipedalism, 313–315
Brave New World (A. Huxley), 333
Brazil, 74–76
Bridgewater, Earl of, 99
British Museum, 170
broad bean, 246
Brontosaurus, 186
brontotheres, 186–188
Broom, Robert, 321
Brown, Roger, 305
Brown University, 40
browsing, 190
Buckland, William, 99–100
Buddhism, 160
Buffon, Comte de (George-Louis Leclerc),
    211, 289
Bumpus, Hermon Carey, 40–41
buttercups, 219
B vitamins, 292

caecilians, 165
Caenorhabditis elegans, 95
calcite, 265
California Academy of Sciences, 37
Calippus, 187–191
Cambrian Period, 19, 25, 269
Cambridge University, 26, 53, 143, 241
"camelopard," 196
Canadian Arctic, 136–139
cancer, 275
Candide (Voltaire), 273–274
Cannon, Annie Jump, 5–6
Capri, Italy, 209
capuchin monkeys, 304
carapace, 156–158
carbamates, 45
carbohydrates, 230–233
carnivora, 113–116
carpometacarpus, 172
Caspian Sea, 64
cassowaries, 81–83

catalysts, 233–234
caterpillars, 101–106
Catholic Church, 20
cattle, 239
"celestial dome," 1, 2, 17
cellulose, 239, 283
centipedes, 54. 57
"central dogma," 254
Cepheid variables, 5–6
Chad, 317
chalk, 169–170
Chambers, Robert, 299
chameleons, 43
chance, in evolution, 226–227
Changeux, Jean-Pierre, 255
Channel Islands, 219
Charles X, 196
cheetahs, 335–336
chelicerata, 57
chemical pathways, poorly designed,
    124–125
cherubim, 151
chimpanzees, 290–293, 297–307, 335–336,
    342
chimp warfare, 304
China, 114–116, 208, 269
chitons, 267
chloroplasts, 236–239
Choeropotamus, 181
chondritic meteorites, 27
Christmas Day (time analogy), 19
cilia, 237
ciliary muscles, 267
Clack, Jenny, 139, 143–145
Cladonema, 262
clams, 267
classification, 87–92
climbing perch, 146
clitoris, 59
cnidaria, 262
coacervates, 233
Coates, Michael, 143
cobalamin, 292
cobras, 160
coccyx, 283–284
cochlea, 58
codon, 243

colobus monkeys, 304
Columbus, Christopher, 192
common cold, 46
compound eye, 263–266
Compsognathus, 171
Confuciusornis, 175–176
Congress, U.S., 170
"contrivances," 108–116
convergent evolution, 49–50, 54, 74–75
Cope, Edward Drinker, 183
Copernicus, Nicholas, 3
coral reefs, 239, 312
cornea, 264–268
Cosmos (Sagan), 14, 27–28
Covington, Syms, 33–36
Cowan, Clive, 241
creation myths, 16–17, 311. *See also* Genesis (biblical book)
Cretaceous Period, 169–174
Crick, Francis, 241–244
crickets, 103
crustacea, 57
cryptodires, 83–84, 157–158
cultural evolution, 335
cup-shaped eye, 262–263
Curie, Marie and Pierre, 26
cuttlefish, 267–268
Cuvier, Baron Georges, 53, 181, 197, 212–214, 289
"C-value paradox," 246
cyanide, 228
cyanobacteria, 236, 238
cyclops, 208
cytochrome c, 93–95, 249
cytosine, 242–243

Da Vinci, Leonardo, 61–62
Daeschler, Ted, 136–139
Dalanistes, 130–131
Dalton, John, 63–64
damselflies, 55
Daouitherium, 220
Daphne Major (Galápagos island), 38
DARC gene, 342
Dart, Raymond, 314
Darwin, Charles, 26, 31–40, 47–49, 53–55, 68, 72–84, 87–92, 97–99, 106–113,

121–127, 170–171, 181–182, 195–200, 225–228, 241, 244–245, 253, 260–263, 269, 274–275, 289–292, 297–300, 342
Davies, Marion, 295
Davis, D. Dwight, 114–115
Dawkins, Richard, 227, 334–335
Dawson, Charles, 315
Dayton, Tennessee, 300
DDT, 45
De Blainville, H. M. D, 181
"deep time," 18
deer, 239, 246
Deinotherium, 213
dementia, 275
Democratic Republic of Congo, 205, 301
Denisovans, 327
Denmark, 141–142
dentary bone, 58–59
Descent of Man, The (Darwin), 274–275, 289–290
design, in nature, 108
desmostylians, 221
deuterium, 228
Devonian Period, 19, 137–145
Dialogues Concerning Natural Religion (Hume), 98, 274
Diamond, Jared, 289, 293
Dicke, Robert, 12
Diderot, Denis, 20
dieldrin, 45
Dimetrodon, 154
Dinohippus, 187–191
dinosaurs, 91, 171–175
distal-less gene, 256
Divine Providence, 210
DNA, 233, 236, 241–250, 291–295; double helix structure, 242
DNA sequencing, 93–95, 291–293
DNA-DNA hybridization, 93–94, 291
Dobzhansky, Thedosius, 342
Döllinger, Ignaz, 64
dolphins, 49–50
"dome of the sky," 2
"Don's Dump Fish Quarry," 154
Doppler effect, 9–11
"double-pulley" astragalus, 133
doublet lens, 265–266

dragonflies, 55–56
Driesch, Hans, 64
dromomerycines, 217
Dubois, Eugene, 68
Dutch East Indies, 68
dwarfed mammoths, 219
dysentery, 239

ear: external, 285–286; muscles of, 286
earliest fossils, 236
East Africa, 302
East Prussia, 64
ecology, 68
ecophenotypic variation, 340
ectoderm, 65
ectopic pregnancies, 279
Edinburgh, Scotland, 20–23
Edops, 154
Egypt, 127–130, 160
Egyptians, 196
Eiffel Tower, 19–20
El Niño, 38
"Elephant Child, The" (Kipling), 207–209
Elephantidae, 217–219
Elephantoidea, 217–219
elephants, 207–222, 275
Elginerpeton, 145–146
Ellesmere Island, 137
elytra, 55
embryo, 58
embryology, 61–71. *See also* Baer, Karl Ernst
    von
embryonic sequence, 68–70
embryonic tail, 283
emu, 81–83
enantiornithes, 175–176
ENCODE project, 248
endemic species, definition of, 77
endoderm, 65
endogenous retroviruses, 247–248, 291
endoparasitism, 101–106
endoplasmic reticulum, 236–237
endosymbionts, 238–239
English sparrow, 40–41
Eocene Epoch, 179, 186
Eohippus, 183, 185, 189
"Eohomo," 185

Eomyticetus, 132
Eozygodon, 216
Epic of Gilgamesh, 72
epigenesis, 62–66
epiglottis, 280–282
Epihippus, 187–189
equidae, 187–194
Equus, 182, 187–194
Eritherium, 220
Eryops, 154
Escherischia coli, 238, 255
esophagus, 280–282
Essay on Man (Pope), 210–211
essential nutrients, 292–293
Estonia, 64–66
Ethiopia, 318
Euglena, 262
eukaryotes, 236–238
Eunotosaurus, 160–161
Eupodophis, 163–165
Eusthenopteron, 140–142, 145–146
evo-devo, 258–260
evolution: of birds, 168–177; chance in,
    226–227; of humans, 311–330. *See also*
    convergent evolution; neutral theory of
    evolution
evolutionary development, 258–260
"evolutionary throwbacks," 176–177, 284
Ewing, Maurice, 241
expanding universe, 11
eye: cornea, 264–268; development of, 258;
    evolution of, 260–269; iris, 269; retina,
    262–269. *See also* compound eye; cup-
    shaped eye; optic nerve; photoreceptors;
    pinhole-camera eye; vertebrate eye
eyeballs, 267–269
eyespots, 262

Fabre, J. H., 101–102
Fallopian tubes, 279
Fallopio, Gabriele, 62
"family tree," 152
family tree of life, 86–95
fatty acids, 230–233
Fayûm beds, Egypt, 213–218, 221
feathers, 174
fibula, 175

fish, 49–50
Fisher, C., 176
"fishing lures," 111–112
Fitzroy, Robert, 35
flagellum, 237–239
Flammarion, Nicholas C., 2
flatworms, 262
flies, 55–56, 255–256
flightless birds, 122
flightless cormorants, 122–124
Flores Island, Indonesia, 327
"Flores man," 327–328
flowers, 99
folate, 292
foramen magnum, 315
formaldehyde, 228, 230
Forster Cooper, Clive, 189
Fossey, Dian, 290, 303
Fox, Sidney, 230–233
Franklin, Benjamin, 21, 209–210
French Revolution, 198, 212
Friend, Peter, 143
"frogamander," 155
frogfish, 146–147
frogs, 84, 152–155
fruit flies, 150, 255–260
Fuller, Harry, 35

Galápagos finches, 32–39, 54, 74–77, 122–124
Galápagos Islands, 33–39, 74–76
Galápagos tortoises, 33–34, 77
Galdikas, Birute, 290, 303
Galileo, 3
Gamow, George, 12
Garden of Eden, 121, 160
Gardner, Allen and Beatrix, 305–307
gastralia, 172
Gaviocetus, 130–131
Gegenbaur, Karl, 67
Gehring, Walter, 258
Geike, Archibald, 22
gel electrophoresis, 92–93, 245
Genesis (biblical book), 72–73, 209, 273. *See also* Adam; Garden of Eden; Mt. Ararat, Turkey; Noah's ark
"genetic assimilation," 254
genetic bottleneck, 335–339

genetic code, 243–244
genetic similarity, 335–336
genetics, 241–250
genotype, 253
genus name, 87
Geoffroy Saint-Hilaire, Étienne, 53, 66, 289
Geographical Distribution of Animals, The (Wallace) 81
geologic time analogies, 17–19
"germ line," 253
Germany, 157. 170
Gerobatrachus, 155–156
giant clams, 239
giant pandas, 113–115
Giardia, 239
Gibraltar, 221
Gingerich, Philip, 127–133
Giraffa, 205–206
giraffes, 195–206, 280
giraffids, 195–206
Gish, Duane, 227
"glass lizards," 165
glasses, 276
glucose, 230–233
goats, 239
gobies, 146
Goethe, Johann Wolfgang von, 53
golden moles, 222
golden rain tree, 45
Golgi bodies, 236–237
Goliath frogs, 154
Gombe Stream National Park, Tanzania, 303
Gomphotheres, 216–218
Gondwana, 81–84, 157
Goodall, Jane, 290, 302–307
goose bumps, 282
gorillas, 290–291, 297–307, 335–336, 342
gorilla sign language, 307
Gould, John, 35–36
Gould, Stephen Jay, 90, 103–107, 110–114, 128, 308–309, 334
Grant, Peter and Rosemary, 38–40
Grant, Robert, 66
Graur, Dan, 248
Gray, Asa, 67, 106

grazing, 190
"Great Chain of Being," 52, 151–152, 210–211, 312
"Great Incognitum," 209–213
Greek mythology, 16, 160
Greenland, 137, 141–143, 157, 337
Greenland Geological Survey, 144
Gregory, T. Ryan, 246
ground sloths, 181
GULO gene, 291
gut flora, 283

Haasiophis, 163–165
Hadrian's Wall, 22–23
Haeckel, Ernst, 67–69, 89–90
hagfish, 268–269
Haldane, J. B. S., 228
halteres, 55–56, 255–256
hammer (malleus bone), 58–59
Harington, Richard, 192
Haringtonohippus, 192
Harvard College, 4–5, 305
"Harvard Computers," 4–6
Harvey, William, 62
haw fly, 40
Hawaii, 78
Hawaiian honeycreeper, 43
hawthorns, 40
hearing aids, 276
heart disease, 275
heartseed vine, 45
heat shock, 254
Hebrews, 72–73
Heezen, Bruce, 241
height, of humans, 340–341
Held, Lewis, Jr., 273, 276, 287
Helicobacter pylori, 337
Helmholtz, Hermann von, 64
Henricksen, Niels, 144
Heraclitus, 3
Herakles, 160
Herbert, Hilary, 169–170
herniated disks, 277
Hertwig, Oscar, 62
Hesperornis, 169–171, 176
heterozygous genes, 341
Hexapoda, 57

Hill, Osman, 302
Himalayas, 341
Hindu mythology, 160, 208
Hipparion, 182–189
hippidions, 192
hippocampus, 297–298
hippopotamus, 132–134
History of Creation, The (Haeckel) 68
HIV, 247, 254
HMS Beagle, 32–38, 54, 74–77, 108, 181, 290
HMS Challenger, 68
Holley, R. W., 243
Holmes, Arthur, 27–28
homeotic genes, 255–260
hominidae, 91
Homo erectus, 68, 87, 322–327
Homo ergaster, 322–323
Homo floresiensis, 327–329
Homo habilis, 87, 322–323
Homo heidelbergensis, 327, 329, 333
Homo rhodesiensis, 327
Homo rudolfensis, 322–323
Homo sapiens, 18, 19, 87, 152, 333
Homogalax, 186–188
homology, 47–60
homozygous genes, 341
homunculus, 62–63
Hooker, Jerry, 189
Hooker, Joseph, 80, 225, 228
horizontal tooth replacement, 219–222
"horned horses," 192–193
horses, 73, 179–189; toes of, 122–123
Hotton, Nick, 155–156
house sparrows, 41
Hox genes, 166, 255–260
Hoyle, Fred, 11, 227
Hubble, Edwin, 6–11
Hubble Space Telescope, 4
Hubby, Jack, 245
human embryo, 70–71
Human Genome Project, 95
human origins, 311–330
human tail, 283–284, 293
Humanson, Milton, 6–11
Hume, David, 20, 98, 274
humor, 264
Hunter, William, 210

hunter-gatherers, 275–276
Hurt, John, 100
Hutton, James, 15–25
Huxley, Aldous, 295, 333
Huxley, Julian, 106–107, 295
Huxley, Leonard, 183
Huxley, Thomas Henry, 106, 168, 170–172, 182–185, 253, 290–299
Hypohippus, 187–190
Hyracotherium, 181–185, 189
hyrax, 181, 222

Ice Ages, 18
ichneumonids, 101–105
Ichthyornis, 169–171, 176
ichthyosaur, 49–50
Ichthyostega, 137–146
idiocracy, 332–333
immune system, 46, 254
immunological distance, 93, 291
incus, 58
Indohyus, 127
Indonesia, 80, 221
industrial melanism, 44–45
influenza, 46
inguinal hernia, 277
"inheritance of acquired characters," 197, 253
introns, 247
Inuits, 341
Ireland, 162
iris, of the eye, 269
iron sulfide, 234
Iroquois mythology, 16–17
island dwarfing, 329
island faunas, 77–79
Isle of Wight, 181
isotopic chemistry, 228
Israel, 165
Italy, 341
ivory, 208

Jacob, François, 255
Janjucetus, 132
Japan, 340
Japanese mythology, 16
Jarvik, Erik, 142–144

"Java man," 68, 322
Jefferson, Thomas, 21, 211–212
Jenny (orangutan), 299–300
Johanson, Don, 318
Joly, John, 25
Judge, Mike, 332
Jukes, Thomas, 244
Julius Caesar, 192, 196
"jumping genes," 254
"junk DNA," 245–250, 287, 291, 334
junkyard, tornado in a, 227
Jupiter, 3
Jurassic Park/Jurassic World (book/movie franchise), 171
Jurassic Period, 19
Just-So Stories (Kipling), 207–208

kakapo, 78
Kalobatippus, 187–190
Kansas, 169–170, 176
Kant, Immanuel, 64
Kaup, Johann Jacob von, 213
kea, 78
Kelvin, Lord (William Thomson), 25–27
Kenya, 302, 317
Kenyapithecus, 318
Khorana, Har Gobind, 243
kiangs, 191–192
kidney beans, 246
Kielmeyer, Karl Friedrich, 65
kiffians, 196
Kimura, Motoo, 244
King, J. L, 244
Kipling, Rudyard, 207–208
Kirby, William, 104–106
kiwi, 78, 81–83
Klasies Mouth Cave, South Africa, 329
knee, injuries to, 279–280
Koch, Lauge, 142
Koko (gorilla), 307–309
Kollar, E. J., 176
Kornbluth, Cyril, 333
Kovalevsky, Vladimir, 182–183
Krakatoa (volcano), 337
Krause, Johannes, 327
krill, 132
Kroehler, Peter, 155

Kushites, 196
Kwan, Alan, 333

La Chapelle aux Saints Cave, France, 325
Lack, David, 36–38
lac operon gene, 255
lactase, 255, 341
lactose intolerance, 341
"ladder of creation," 151, 186
"ladder of life," 312
Laetoli, Tanzania, 318
Lake Moeris, Egypt, 214
Lake Turkana, Kenya, 321
Lamarck, Jean Baptiste de Monet,
    Chevalier de, 196–198, 226, 289–290
"Lamarckian inheritance," 198, 253
Lambdotherium, 186–188
Lamont-Doherty Geological Observatory,
    241
Lampsilis, 111–112
Laplanders, 341
lateral lines, 142
"law of correlation of parts," 212
lead, 27
Leakey, Louis S. B., 302–303
Leakey, Mary, 302, 321
Leakey, Richard, 322
"Leakey's Angels," 303. *See also* Fossey,
    Dian; Galdikas, Birute; Goodall, Jane
Leavitt, Henrietta Swan, 5–6
Lebanon, 165
Leeuwenhoek, Antonie von, 62
left recurrent laryngeal nerve, 201–204,
    280–281
Leibniz, Gottfried Wilhelm, 273–274
Leidy, Joseph, 182–183
Lemaître, Georges, 11
lens alpha crystallin, 95, 264–268
Lents, Nathan, 279, 288
leprosy, 342
Lernean Hydra, 160
lesser panda, 115–116
Lewis, Edward B., 255
Lewis and Clark expedition, 212
Lewontin, Richard, 245
Liang Bua Cave, Indonesia, 327
Liaoning Province, China, 174

Lightfoot, John, 17
limes, 292
limpets, 267
LINEs, 247
Linnaean Society, 80
Linnaeus, Carolus, 72, 73, 86–92, 126, 289,
    297
lipid bilayer, 231–232, 235
lipids, 230–233
Llanocetus, 132
lobed fins, 139–141
Loberg Lake, Alaska, 42
Locke, John, 20–21
London Clay, UK, 179
London Zoo, 299
"Lucy," 318
Lukeino Formation, 317
lungfish, 84, 90–91, 141, 246
Lusk, Wyoming, 190
Lyell, Charles, 80, 104
lysine, 243
lysorophids, 166

macrocytic anemia, 292
macroevolution, 150–166, 255–260
Madagascar, 77–78, 156, 176, 196
maggots, 226
Malaku Islands, 80
malaria, 80, 342
Malay Archipelago, 80–82
Malaysia, 80
malleus, 58
"mammal-like reptiles," 58
Mammalodon, 132
mammoths, 213. *See also* dwarfed
    mammoths
Mammut, 216–217
Mammuthus, 217
mammutidae, 216–218
manatees, 221–222
mandrills, 304
"march of progress," 313
Margoliash, Emanuel, 249
Margulis, Lynn, 237–239
Marsh, Othniel Charles, 169–171, 183, 193
marsupials, 74–75, 81–83
mastodons, 209–211, 216–218

Matthaei, Heinrich, 243
Matthew, William Diller, 179–182
Maury, Antonia, 5
maxillary sinus, 277–278
Mayr, Ernst, 258, 334
McClintock, Barbara, 254
McKenna, Malcolm, 221
McPhee, John, 15, 18
Meckel, Johann Friedrich, 65, 68
Megahippus, 187–190
Megalonyx, 212
Megalosaurus, 99
meme, 335
Mendelian inheritance, 197–198
Mendi, Joe (chimpanzee), 300
Merschowski, Konstantin, 237
Merychippus, 187–191
mesoderm, 65
Mesohippus, 184–190
Mesolithic Period, 339
mesonychids, 127–128
mesotarsal joint, 172–175
Mesozoic Era, 18
Metamorphoses, 3
Metaxygnathus, 145–146
methane, 228–229
microhylidae, 84
microscope, 61–71
microwave background, of universe, 11–13
mid-ocean ridges, 234–235
middle ear, 58
Middle East, 72–73
milk consumption, 341
Milky Way, 6
Miller, Stanley, 228–230, 241
Miller-Urey experiment, 228–230
millipedes, 54, 57, 258
Mindell, David, 32
Minippus, 189
Miocene Epoch, 202, 213
miohippus, 184–190
"missing links," 150–166, 300, 312
mites, 100
mitochondria, 236–239. *See also*
    mitochondrial DNA
mitochondrial DNA, 95, 191–192. *See also*
    mitochondria

Mivart, St. George Jackson, 261
moas, 78
mockingbirds, 33–35, 77, 99
Moeritherium, 214–216, 219–220
moles, 48–49
molecular clock, 249
molecular phylogeny, 92–96, 116. 132–134,
    191
molluscs, 265–267, 312
molting, 264
Mongolia, 186
"monkey to man" progression, 152–153, 313
"monkey with a typewriter" analogy, 227
Monod, Jacques, 255
Monte Verde, Chile, 339
Montesquieu, Charles-Louis Secondat,
    Baron de, 20
Moores, Eldridge, 15
moray eels, 146
mosquitoes, 45, 55
mosquito fish, 43
Mount Wilson Observatory, 6–11
mouse, genome of, 247
Mousterian culture, 325
Moyne, Charles le, 209
"Mrs. Ples," 321
Mt. Ararat, Turkey, 72, 78, 81
mudskippers, 146–147
Müller, Johannes, 66
mummified mammoths, 219–220
Museum für Naturkunde, Berlin, 170
musk deer, 217
mustangs, 192
Myriapoda, 57
Mysticetes, 131–132

Najash, 165
Nannippus, 187–191
Napier, John, 302
"Nariokotome Boy," 322–323
nasal sinuses, 277–278
Natatanuran frogs, 84
National Geographic (magazine), 301–302
Native Americans, 336–337
Natural History Museum, London, 35
Natural History Museum of Paris, 198
natural selection, 80, 198–199, 225–226

Natural Theology (Paley), 97, 108–112, 260, 274

Naturphilosophie, 53, 65, 67, 97

nautilus, 264, 268

Neanderthals, 323–326

Nebraska, 191

"necking," 200

neoteny, 294–295

neuronal development, 293

neutral theory of evolution, 244–246, 249

neutrinos, 241

New Year's Eve (time analogy), 19

New Zealand, 78

Newman, Horatio, 283

Newton, Isaac, 4, 20

newts, 140

Nicholson, John, 143

nictitating membrane, 287

Nierenberg, Marshall, 243

Nietzsche, Friedrich, 332–333

nipples, male, 274, 287

Nisir, 72–73

Noah's ark, 72–74, 81

noncoding DNA, 245–250, 291

Norway, 141–142

nuclear DNA, 95, 191–192

nucleotides, 242–244

nucleus, 236

Numidotherium, 220–221

"Nutcracker Man," 321

Ochoa, Severo, 243

octopus, 267–269

odontocetes, 131–132

Odontochelys, 158–161

odontornithes, 169–171

oil, 230–232

okapi, 203–205

Okapi Wildlife Reserve, 205

Oldowan technology, 322

Olduvai Gorge, Tanzania, 303, 321

olfactory region, 292

On the Origin of Species (Darwin), 32, 46, 54, 61, 67, 72, 76, 109, 126, 170–171, 182, 244, 260, 274, 290, 297

onagers, 191–192

"one gene, one protein" dogma, 250

"one species at a time," 315–316

"onion test," 246

ontogeny, 61–71

"ontogeny recapitulates phylogeny," 68

onychophoran, 264

Oparin, A. I., 228

ophidiophobia, 161

optic nerve, 267–269, 276

orangutans, 299–300, 335–336

orchids, 109–111

organelles, 236–238

organophosphates, 45

origin of life, 225–240

Orohippus, 184–189

Orrorin, 87, 317–318

ossicones, 203

Osteolepis, 142

ostrich, 81–83, 280

Ostrom, John, 172

"Out of Africa" migration model, 327, 338

ovaries, 59, 279

ovipositor, 58, 101

ovism, 62–66

Owen, Richard, 35, 53, 170, 179–182, 297–299

oxygen, atmospheric, 226

Pääbo, Svante, 327

Pachyrachis, 165

paddlefish, 148

pain, problem of, 100–101

Pakicetus, 127–130

Pakistan, 127–130, 133

Palaeomastodon, 214–216

Palaeotherium, 182–183, 189

Paleocene Epoch, 186

Paley, Rev. William, 97–99, 108–111, 274

Pallas, Peter Simon, 212

palmar grasp reflex, 287

pan-selectionism, 244, 248–250

panda's thumb, 113–116

Pander, Hans Christian, 65

Panderichthys, 145–146

Pangea, 19

Pappochelys, 160–161

parabasalids, 239

Parahippus, 187–190

Paranthropus, 87, 302, 320–322
Paranthropus boisei, 320–321
Paranthropus robustus, 320–322
Parathropus aethiopicus, 320–321
"parson's nose," 172
Pasteur, Louis, 226
Patagopteryx, 176
"pattern of unification," 65
Patterson, Francine, 307–308
Pauling, Linus, 248
Pax-6 gene, 258, 262, 285
PCR. *See* polymerase chain reaction
Pearl Harbor Day (time analogy), 19
Peebles, Jim, 12
"Peking Man," 324
pellagra, 292
Pelomyxa, 239
penguins, 49–50
penis, 58
Pennsylvania, 137
Penzias, Arno, 11–13
peppered moth, 44–45
perissodactyl, 182
peritoneal membrane, 277
periwinkles, 43
Permian Period, 19, 154
pesticide resistance, 45–46
phenotype, 254
phenylalanine, 243
Philosophe Zoologique (Lamarck), 289–290
philosophical optimism, 273
Phiomia, 214–216
Phosphatherium, 220
photoreceptors, 258, 260–269
phylogeny, 68, 86–95
phylum, 68
"Piltdown Man," 241, 315
pinhole-camera eye, 263–268
"Pisces" (taxonomical group), 90
Pithecanthropus erectus, 68, 322
placentals, 74–75
planarians, 262
plastron, 156–158
Plato, 47
Pleasure Point, Washington, 41–42
pleiotropic linkage, 250
pleiotropy, 250

pleurodires, 83–84, 157–158
Pliohippus, 184–191
Pliolophus, 189
polar bears, 136–137
polymerase chain reaction (PCA), 95
polymerization, 230–233
Polypterus, 148
pongidae, 91
Pope, Alexander, 210–211
Powell, John Wesley, 168–169
Precambrian, 18
preformationism, 62–65
premature birth, 279
Presbyterian Church, 20
Prezewalski's horse, 192
primates, 289
Primelephas, 217
"primordial soup," 228
Principles of Geology (Lyell), 104
probability argument, 227–228
proboscideans, 207–222
Proconsul, 302
Proganochelys, 157–158
Project Echo, 11
prokaryotes, 236–238
Prolibytherium, 203
proline, 243
prostate cancer, 278–279
prostate gland, 278–279
prosthetics, 276
proteinoids, 233
proteins, 230–233
Protista, 68, 262
protoceratids, 217
"protolife," 233
protomammals, 58
protorohippus, 187–189
protozoans, 262
"pseudelephant," 210
pseudogenes, 247, 287, 291, 292
pterosaurs, 49–51, 172, 174
puffer fish, 246
"pure" research, 13
purple nonsulfur bacteria, 238
pygostyle, 172–177
pyrethroids, 45
pyridoxine, 292

pyrite, 234–235
Pythagoras, 61

qilin, 196
quadrate bone, 58–59
quagga, 191

rabbits, 43
raccoons, 116
"races," human, 339
radial sesamoid bone, 113–116
Radinsky, Leonard, 189
Radinskya, 189
radioactive decay, 27
radiolarians, 68
radiometric dating, 27
Rahonavis, 174–176
ratites, 81–83
ray-finned fish, 145–147
Ray, John, 108
recapitulation, 65
recessive genes, 341
red panda. *See* lesser panda
red shift, 9–11
Rees, Martin, 333
refractive index, 264–265
regulatory genes, 292
Reines, Frederick, 241
Reisz, Robert, 155–156
repetitive DNA, 247
Reptilia, 91
Republic of Congo, 205
Republic of Georgia, 324
rete mirabile, 200
retina, 262–269
retroviruses, 254
rheas, 81–83
rhinos, 182, 186–188, 217
Richard, William, 179
RNA, 235, 243
Rocky Mountains, 183
Rodhocetus, 128–130
Romania, 324
Romans, 22, 196
roundworms, 100, 246
Rousseau, Jean-Jacques, 20
rudimentary organs, 121–134, 274–287

Rusinga Island, Kenya, 302
Russia, 64, 212
Rutherford, Ernest, 26–27

Sagan, Carl, 14, 27–28
Sahelanthropus, 87, 316–317
salamanders, 140, 154, 246
Salk, Jonas, 241
salmon, 275. *See also* sock-eye salmon
Samotherium, 205
San people, 195
Santa Barbara, California, 219
Sapolski, Robert, 292–293
sarcodes, 261–262
sarcopterygii, 91
Save-Söderbergh, Gunnar, 142
Scala naturae, 151–152, 312
scallops, 267
Schaller, George, 297–303
Schwab, Ivan, 261
sciatic nerve, 280
science fiction, future humans in, 332–333
scoliosis, 277
Scopes Monkey Trials, 300–301
Scotland, 20–22
Scottish Enlightenment, 20–22
scrotum, 277
sculpins, 146
scurvy, 292
sea anemones, 262
sea jellies, 68, 262
sea urchins, 64
seafloor mapping, 241
semen, 62
semilunate carpal, 172–174
serial homology, 54
Serres, Étienne, 65, 68
sexual selection, 42
Shakespeare, William, 227
Shanidar Cave, Iraq, 325
sharks, 269
Sherpas, 341
Shiva, 160
shovel-tuskers, 217
Shubin, Neil, 135–148, 189–190
Siberia, Russia, 192

Siberian mammoths, 211
Siccar Point, Scotland, 24–25
sickle-cell anemia, 341–342
side toes, 192–193
side-necked turtles, 83–84
Sifrhippus, 189
simian AIDS, 293
SINEs, 247
sinus headaches, 277
sirenian, 221-222
Sivatherium, 203
sixth gill arch, 201–204, 280
skin pigmentation, 339
skinks, 165
Slipher, Vesto, 6–11
Slovenia, 162
"slow worms," 165
Smith, Adam, 20
snails, 262–269
snakes, 160–166; hipbones of, 122–123
soapberry bug, 45
sock-eye salmon, 41–43
sodium, 6
Solounias, Nikos, 205
"soma," 253
somites, 54–57
South America, 192
South Dakota, 182, 189
species name, 87
Species Plantarum (Linnaeus), 87
spectacles, 273
spectral analysis, 6–11
spermism, 61–66
spherical aberration, 265
spiders, 57
spirochaete, 238
Spitsbergen, 137
sponges, 312
spontaneous generation, 225–227
squamosal bone, 58–59
Standen, Emily, 148
stapes, 58
starches, 230–233
"steady-state" model, of universe, 11
Stensiö Berg, Greenland, 143
Stetten, DeWitt, 243

sticklebacks, 42–43
stingers, 58
stirrup (stapes bone), 58–59
streptomycin, 238
strokes, 275
structural genes, 292
sturgeon, 148
Sues, Hans-Dieter, 160
sugars, 230–233
Sulloway, Frank, 33–35
Sumerian mythology, 72–73
Sundman, John, 245–246
symbiosis, 237–239
synapsids, 58
synsacrum, 175
syphilis, 238
Systema Naturae, Regnum Animale
  (Linnaeus), 87

tailbone, 283–284
Takracetus, 130–131
Tambora (volcano), 337
tapeworms, 100
tapirs, 182, 186–188
tarsometatarsus, 176
Tasmania, Australia, 285
"Taung child," 314
taxonomy, 87–92
Tbx genes, 166
telescopes, 3–4
Telstar Satellite, 11
tenrecs, 222
termites, 239
Ternate, Indonesia, 80
testes, 58; descent of the, 277
testosterone, 59
Tethys Seaway, 221
Tethytheria, 221
tetracycline, 238
tetrapods, 91, 137–148; forelimb of,
  48–49
Texas fossil beds, 154
Thailand, 157
Tharp, Marie, 241
Theorie de la Terre (Buffon), 211
Theory of the Earth (Hutton), 22

theropod dinosaurs, 172
Thewissen, Hans, 128–130, 133
thiamine, 292
"third chimpanzee," humans as, 293
thumb, 250. *See also* panda's thumb
Tiktaalik, 137–145
time: concepts of, 15–18; geologic, 15–21
Time Machine, The (Wells), 333
tinamous, 81–83
Toba catastrophe, 337–339
Toba (volcano), 337–338
"toe off" walking, 318. *See also* bipedal
  posture; upright walking
tools, use of, 304–305
toothed whales, 131–132
"Toumai," 317–318
trachea, 280–282
transfer RNA, 243
transitional fossils, 150–166
transposons, 247
"tree of life," 152–155
Triadobatrachus, 156
Triassic Period, 157–158
trilobites, 57, 265–266
tuberculosis, 292–293, 342
Tugen Hills, Kenya, 317
Turing, Alan, 241
Turkey, 72, 81
turtles, 156–161. *See also* side-necked
  turtles
Twain, Mark, 19–20

Übermensch, 332–333
ulcers, 337
Uniformitarianism, 22
Universal Deluge, 17
University of Dorpat, 64, 67
University of Edinburgh, 27, 53, 66
upright posture, 276–280
uracil, 242–243
uranium, 27
urethra, 278–279
Urey, Harold, 228–230
urination, 278–279
U.S. Geological Survey, 168–169
Ussher, James, 17

Utnapishtim, 72–73
UV radiation, 264, 339

Valentine Formation, Nebraska, 191
Van Valen, Leigh, 127
vas deferens, 279
Velociraptor, 91, 172, 174
velvet worms, 264
Ventastega, 145–146
Venter, Craig, 95
Venus, phases of, 4
vertebrate eye, 267–269, 276
vestigial organs, 121–134, 274–287, 334
Viens, Jo, 300
Vieraella, 156
Vietnam, 208
Virunga volcanoes, 301, 303
viruses, 46, 150
Vishnu, 160
vitamin C, 291–292
vitamin D, 339
voice box, 201–204, 280–281
Voltaire, 20, 273–274
vomeronasal organ, 287
Von Humboldt, Alexander, 79
Vorona, 176
Voyage of the Beagle (Darwin), 33, 37, 76

Waddington, Conrad, 254
Walcott, Charles Doolittle, 170
Walker, Alan, 321
"walking catfish," 146–147
Wallace, Alfred Russel, 74–84
Wallace's line, 81–82
Walsh, Benjamin, 40
"warm little pond," 225, 228, 234
Washoe (chimpanzee), 305–309
wasps, 58
"watchmaker" analogy, 97–98
watercress, 246
Watson, James, 241, 254
Watt, James, 20
weasels, 116
Weiner, Jonathan, 31, 39
Weissman, August, 253
Wells, H. G., 333

whales, 48–49, 125–134; hipbones of, 123–130.
  See also baleen whales; toothed whales
Whitehead, Alfred North, 47
Wiedersheim, Robert, 282–283
Wilberforce, Samuel, 298–299
Wilkinson, David, 12
Wilson, Robert W., 11–13
wisdom teeth, 284–285, 340
Wolff, Caspar Friedrich, 63, 65
Woodburne, Mike, 186
Woodward, John, 17
woolly rhinoceros, 212–213

World War II, 340
worldwide flood myth, 72–73

Y chromosome, 59

Zarafa, 196
zebras, 191–192
Zhoukoudian caves, China, 324
Zinjanthropus boisei, 302, 320–321
Ziusudra, 72
Zuckerkandl, Émile, 248
Zygolophodon, 216